Biblical Astrology
A Biblical Model to Astrology

By

DANIEL G. McCRILLIS, Th. D.
Pastor, Gateway Baptist Church in St. Louis Missouri

Gateway International Publishing
St. Louis Missouri

All Scripture is given from the
King James Version

Library of Congress Control Number: 2012932928
Printed and Made in the United States of America

"And they that be wise shall shine as the brightness of the firmament; and they that turn many to righteousness as the stars for ever and ever."
(Daniel 12:3)

Table of Contents

Introduction

Considering the heavens (Ps. 8:3) is one of the most fascinating challenges the Bible presents to humanity. In light of man's existence, the stars puzzle the human psyche as to our being and purpose. To most people, they are just little orbs with no real significance to life or the Scriptures; however, those of us who have accepted the Bible's challenge know differently.

From the human standpoint, one can find the trademarks of a Master Architect in the smallest cell to the large blue whale, yet out of the thousands of stars visible to the naked eye, one may wonder whether there is any *design* and *architecture*. The stars appear to be un-uniform and chaotic, but when correlated with the Word of God and sectioned off into their decans (secondary constellations to the main twelve), one cannot help but to arrive with *cosmos* (order) out of the seeming *chaos* (disorder).

Since the stars are a larger part of God's creation, and can be seen by man, then philosophically we need to ask, "What purpose do the stars have in relation to humanity?" Are we supposed to visit their exoplanets through space exploration, scientifically deduce their compounds with astronomical spectroscopy, or figure their distance using stellar parallax?

In light of man spending billions of dollars in research trying to find answers to our existence: who created us?, where are we going when we die?, are we alone in the universe?, and what is life really about?, *Biblical Astrology* poses a more practical approach to be taken into consideration. In this work, we will avoid the magnitudes, observations, order and mathematical calculations of Astrometry; focusing – for the most part – on four basic theological principles.

1) All the Stars Have Names

The first principal is that every star carries a name given by God (Psa. 147:4; Isa. 40:26).[1] These names all have interpretations which express

---

[1] He telleth the number of the stars; he calleth them all by *their* names (Ps. 147:4). Lift up your eyes on high, and behold who hath created these *things*, that bringeth out their host by number: he calleth them all by names by the greatness of his might, for that *he is* strong in power; not one faileth. (Is. 40:26)

teachings revealed in the Bible. Significantly, there is no confusion between the stars' and Scripture's message, they are one and the same (1ˢᵗ Pet. 1:20-21; Ps. 19:1-6).[2] If one wishes to know the signs and stories of the stars, he must study their names. Once this is done, he can see that the *heavenly signs* are a by-product of higher Intelligence, which the Bible claims belongs to the Creator.

2) The Stars Teach a Sign

The second principal is that stars teach a sign (Gen. 1:14). These images or signs are representative of the stars' names, not the stars' names to an image. That means one shouldn't look up in the night-sky to imagine an image and then assign names to the stars that support the image, but vice-versa. This is important in that it takes the *mystic* approach out of Astrology and gives us one that is *natural,* allowing the intelligence that God placed behind the stars to speak to ours. Francis Wayland, a powerful theological mind from two centuries ago said:

> We are frequently referred in theological writings to the works of creation, as a proof of his greatness and wisdom; and the remark has been made, not without reason, that 'the stars teach as well as shine.' The discoveries of modern astronomy not only assure us, that there is a God, but impart this additional assurance, that he is above all others, to whom the attributes of divinity may have been at any time ascribed.[3]

3) The Stars Tell a Story

The third principal is that the stars tell stories. Jeremiah 10:12 informs u that God "stretched out the heavens by his discretion." The word *discretion* is the Hebrew word *ta-wboon,* which means intelligence. By associating this intelligence factor within the astronomical record a most divine message begins to appear. In Amos 9:6, the prophet declared that "it is he that buildeth his stories in the heaven." These stories are

---

[2] Forasmuch as ye know that ye were not redeemed with corruptible things, *as* silver and gold, from your vain conversation *received* by tradition from your fathers; But with the precious blood of Christ, as of a lamb without blemish and without spot: Who verily was foreordained before the foundation of the world, but was manifest in these last times for you, Who by him do believe in God, that raised him up from the dead, and gave him glory; that your faith and hope might be in God. (1ˢᵗ Pet. 1:18-21)
[3] Francis Wayland, *Elements of Intellectual Philosophy,* Published by William Hyde, Brunswick 1827.

successive chapters in the constellations which unveil God's plan for the ages, time's riddle, man's purpose, and the great conflict between good and Evil.

Historically, these stories were known to the Bible writers, beginning with Virgo and her decans, ending with Leo and his. Consider the following astrological facts known just to Job: (a) "Orion's bands" (Job 38:31), (b) "Pleiades' sweet influences" (Job 38:31), (c) "Mazzaroth's seasons (Job 38:32)," (d) Arcturus' sons (Job 38:32), and (e) "Draco's crookedness" (Job 26:13). All of these examples shed light on *the stories* found within the stars. Philo said of the stars:

> The air is the abode of incorporeal souls, since it seemed good to their Maker to fill all parts of the universe with living beings. He set land-animals on the earth, aquatic creatures in the seas and rivers, and in heaven the stars, each of which is said to be not a living creature only but mind of the purest kind through and through.[4]

4) The Stars Declare the Glory of God

The fourth principal that we must recognize is that the stars declare the glory of God (Psa. 19:1). Let's look at Psalms 19:1-5 for a moment,

> The heavens declare the glory of God; and the firmament sheweth his handywork. Day unto day uttereth speech, and night unto night sheweth knowledge. *There is* no speech nor language, *where* their voice is not heard. Their line is gone out through all the earth, and their words to the end of the world. In them hath he set a tabernacle for the sun, Which *is* as a bridegroom coming out of his chamber, *and* rejoiceth as a strong man to run a race.

This reference reveal that the stars declare God's glory by: (a) proclaiming God's handiwork, (b) showing knowledge, (c) speaking a message, and (d) setting a tabernacle for the sun. The understanding of the Zodiac being wrapped around the sun in twelve impeccable sections

---

[4]Classical Loeb Library, Philo V, On Dreams, Lin. 135, Translated by F.H. Colson and G.H. Whitaker, 1934.

with a precise percentage of 8.333 was the main reason why primitive man worshipped astrology. From this phenomenal pantheon, early civilization attributed several deities, while overlooking the Supreme Deity.

Without the Zodiac it would be impossible for the stars to "be for signs, and for seasons, and for days, and years" (Gen. 1:14). This provided the world with a calander year.

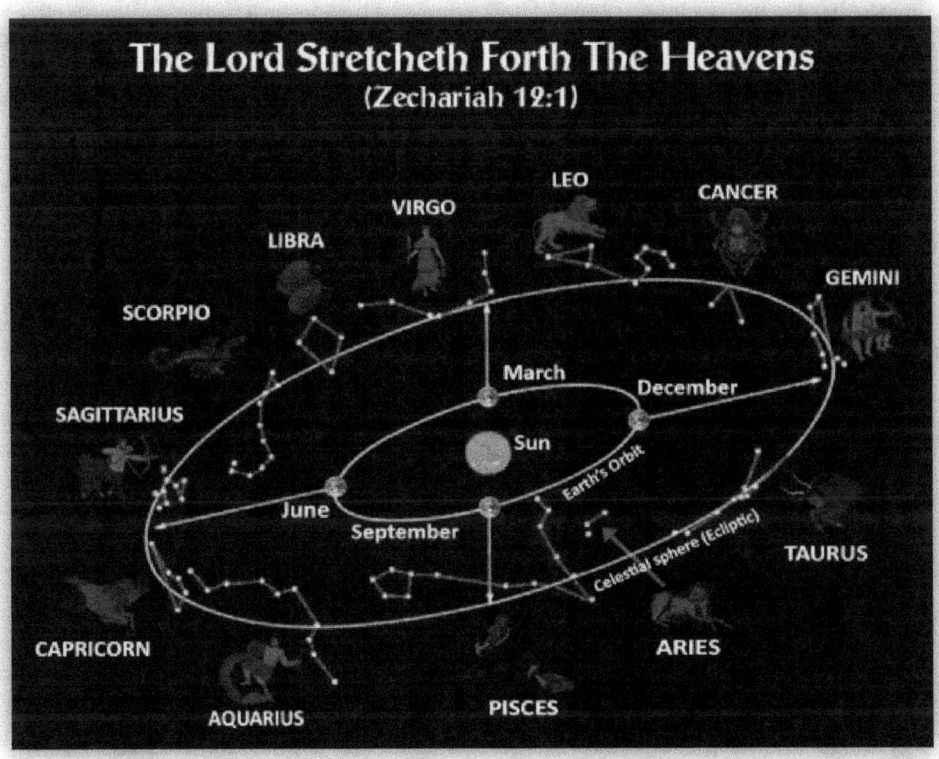

The most primitive civilization in Egypt knew there was Divine Intelligence behind the Cosmos. They believed that their life, fate, and personality were linked to the astrological signs. In fact every nation on earth has deep roots with the stars. These secret teachings must have come from the antediluvian era and the Tower of Babel.[5] This astronomical significance is found throughout Scripture with the blessings given to the sons of Jacob (Gen. 49) being patterned after the

---

[5] *Biblical Egyptology*, Chap. 2 The Ancients and the Tower of Confusion, sec. Cosmological Significance, Dan McCrillis, Gateway International Publishing, 2018.

12 Constellations, and also with Moses' prophecy in Deuteronomy 33 and the tabernacle encampment in Numbers 2, the astronomical teaching of David (Ps. 19), and the Temple encampment by Solomon (1st Chron. 27).

Some of these secrets unveiled Israel's conflict to bring about the Messiah, which catapulted a new dimension of believers to be grafted into Israel's divine olive tree (Rom. 11). In light of Biblical Astrology, we will discover that the stars proclaim the most thorough pronouncement of the Gospel.

Biblical Astrologers

Several Bible characters stepped out of the status quo of being a star-gazer and graduated to bona-fide students of the stars. Consider the following men: Jacob, Joseph, Moses, Job, David, Isaiah, Jeremiah, Amos, Ezekiel, the wise men, Paul, and Jesus Christ himself. As irrelevant as most people believe the stars to be, it is astounding that God would be so adamant to keep their purposes alive in his Word. For instance think about their following uses:

a) To bring about prophecy (Num. 24:17)
b) To declare his glory (Ps. 19:1)
c) To identify himself as One (Rev. 22:16)
d) To pattern their stories and truths after some of the most vital doctrines in the Bible? (Amos 9:6)

Star Doctrine

The stars magnify many doctrines surrounding the Soteriology that are vital to every believer. In Virgo, we see the importance of *the Incarnation*, fulfillment of the Old Testament prophecy, *the Virgin Birth*, and God in the person of Jesus Christ. In Libra, we see the importance of *the Justification of the Believer*, *Atonement*, the finished work of Calvary, *the Vicarious Death of Christ*, and *the Imputation* of Christ's benefits to the believer in light of Justification. In Draco, we see the importance of the doctrine of *a Personal Devil*, and his crafty plot to disrupt the plan of God. In Leo, we see *Eschatology* in the stars. In Bootes, we see Christ as the Great Shepherd and his multi-task of reaping the harvest and tending the sheep. In Orion, we see Christ as the glorious fulfillment of the Old Testament's *Coming Branch* prophecies and the Devil's defeat from Calvary. Stories like these will be expounded throughout this book.

The Field of Biblical Astrology

In the past two hundred years there have been merely a few handfuls of men and women that have succeeded in writing works on biblical astrology. Every Christian who ventures into this field of study should know three paramount works: *Mazzaroth* by Frances Rolleston, *The Gospel in the Stars* by Joseph Seiss, and *the Witness of the Stars* by E.W. Bullinger. Frances Rolleston's work called *Mazzaroth* laid the foundation, reaching back into history and accumulating a lot of factual importance and biblical relevance. *Mazzaroth*, also listed and documented every main star in Arabic, Greek, Hebrew and Coptic for every constellation and decan. Her work is invaluable to the field.

In 1882 Joseph Seiss wrote a fascinating work called *the Gospel in the Stars*. In this book he walks us completely through the Zodiac and all of the constellations' decans. He put on the bottom shelf, the relevance of the stars and added some important knowledge to Rolleston's work.

In 1893 E.W. Bullinger wrote, *the Witness in the Stars*. He added more of a dispensational view to Biblical Astrology. Although a lot of his work and thought is that of Seiss, he discovered much of his own material and brought in many Hebrew connections, thought, and direction.

Our Difference

These three works are *the classics* for the field of *Biblical Astrology*. Most of the recent works on the stars have been found to be condensed works almost completely in thought and lesson of the former three. In our work there are zodiacal signs where we freelanced from the predecessors. In some constellations, we felt most writers just patterned their intelligence after Rolleston, Seiss, and Bullinger. Although the successor's thoughts are scriptural, the interpretation, we felt, was not after the proper model laid out in this introduction. In these instances, we explain thoroughly why we went another direction.

There were several places where the successors did not use the intelligence behind the star names to provide proper star doctrine, but followed the mythical teachings of the Greek era. For instance, most followed the dual nature of Centaurus – the third decan in Virgo – to depict Christ, when the stars seem to depict the dual nature of Satan who can transform into an angel of light (2nd Cor. 11:14). The images revealed by the stars show Centaurus trampling and piercing to death *the Victim* Lupus (a picture of Christ). Another instance is found in the second decan in Scorpio, Ophiucus. Most predecessors followed Seiss' Great Physician analogy, when not a single star name in the constellation depicts such a teaching.

Furthermore, in our model we followed the more ancient pictorials from the zodiacs of Dendera and Esnah in Egypt over the Ptolemaic models; whereas, some of the former works followed the Greek mythological teachings and images. For instance the Ptolemaic Model has *the crab* for Cancer instead of *the scarab beetle.* Most had *the Hair of Bernice* for Coma, whereas we have the *Desired Son of Virgo.* In addition, and due to our technological era we were able to use real images of the stars themselves to make more exact and modest image-work for the constellations.

Extra Credit

The Bible informs us that the stars sing (Job 38:7; Psalm 19:4; and Rom. 10:18). If the stars are able to sing, then they must carry a musical note. Paul, quoting the Septuagint's Psalm 19:4 in Romans 10:18, said, "But I say, Have they not heard? Yes verily, their **sound** went into all the earth, and their **words** unto the ends of the world." As mentioned previously, Paul was referring to the stars. The word *sound* is the Greek word φθογγος, pronounced in English as **f-thong-os,** and means "a musical note." As the earth journeys along its ecliptic, it travels through the Zodiac. Each month the earth—being in position from the sun—is housed in one of the constellations. This gives us signs, seasons, days, months and years (Gen. 1:14). If the stars in each of those constellations were in a musical graph, starting with c-note and having 18 notes successively; as the earth would travel on its ecliptic, each star would become a specific **kind** of musical note for the sun. It would then be up to man to determine what **type** of note each star would be (whether half, whole, quarter, eighth, etc.). This is how we came up with music for the constellations. God created music, musical notes, stars, constellations, the earth and its ecliptic; we just found a way to combine them in a way which allows the stars to do what the Bible says they can.

So, found within the pages of this book are several musical pieces coming from the stars themselves as a beautiful testament to God's intelligent design.

A Source of Faith

The teachings found within the stars are also a relevant source for faith. This source is in the message or gospel. The apostle Paul quoted the Septuagint's Psalm 19:4 in Romans 10:17-18 when he said, "So then faith cometh by hearing, and hearing by the **word** of God. But I say, Have they not heard? Yes verily, their sound went into all the earth, and their **words** unto the ends of the world." If you check Paul's Old Testament quote, you will find that he was contextually referring to the celestial bodies.

Psalm 19:4, εἰς (into) πᾶσαν (all) τὴν (the) γῆν (earth) ἐξῆλθεν (spread abroad) ὁ (the) φθόγγος (sound) αὐτῶν (their), καὶ (and) εἰς (into) τὰ (the) πέρατα (ends) τῆς (of the) οἰκουμένης (inhabited earth) τὰ (the) <u>ῥήματα</u> (words) αὐτῶν. (their)[6]

Romans 10:17-18, Ἄρα (so then) ἡ (the) πιστις (faith) ἐξ (cometh by) ἀκοῆς, (hearing) ἡ_δε (and the) ἀκοη (hearing) δια (through) <u>ῥηματος</u> (word) θεοῦ (God). ἀλλα (contrarywise) λεγω (I say), Μη οὐκ_ (have they not) ἤκουσαν (all heard); μενοῦν γε (yes verily) εἰς (into) πᾶσαν (all) την (the) γῆν (earth) ἐξῆλθεν (spread abroad) ὁ_ (the) φθογγος (sound) _αὐτῶν (their), και (and) εἰς (into) τα (the) περατα (ends) τῆς (of the) οἰκουμενης (inhabited earth) τα (the) <u>ῥηματα</u> (words) αὐτῶν (their).[7]

There are two underlying Greek words for the one English word *word*. One is *logos*, which means "reasoning" or "logic." Thus, Christ is the ο (the) λογος (Logic) in John 1:1, referring to the Intelligence behind the creation. *Logos* can also refer to a word (Luke 12:10) or book (Mark 7:13 and Hebrews 4:12). Another underlying Greek word for *word* is ρημα, pronounced *rhay-ma,* which means "sayings" or "topic of discussion."[8] It is significant to notice above that "word" in Psalm 19:4 and Romans

---

[6] *The New Acrostic Study Bible*, Volume III, Poetical Books, Gateway International Publishing, 2016
[7] Ibid. Vol. V.
[8] *Strong's Exhaustive Concordance*, James Strong, Hendrickson Publishers, Peabody, MA.

10:17-18 are the underlying Greek "ρημα," not "λογος" and should be interpreted as "a specific topic of discussion."

Conclusion: Through study, faith can be obtained by the messages given within the word of God, and also the original Astrological record recorded in the stars. However, it is important that we follow the Bible's guidelines; namely, the four basic principals in this introduction.

Although the stars are not necessary to know any of the major teachings of the Bible, one finds through study that it is not the case with history. Therefore, it is no surprise why we find the stars as an excellent companion to good doctrine.

Daniel McCrillis, Th. D.

VIRGO "THE VIRGIN"
CHAPTER 1 SECTION 1

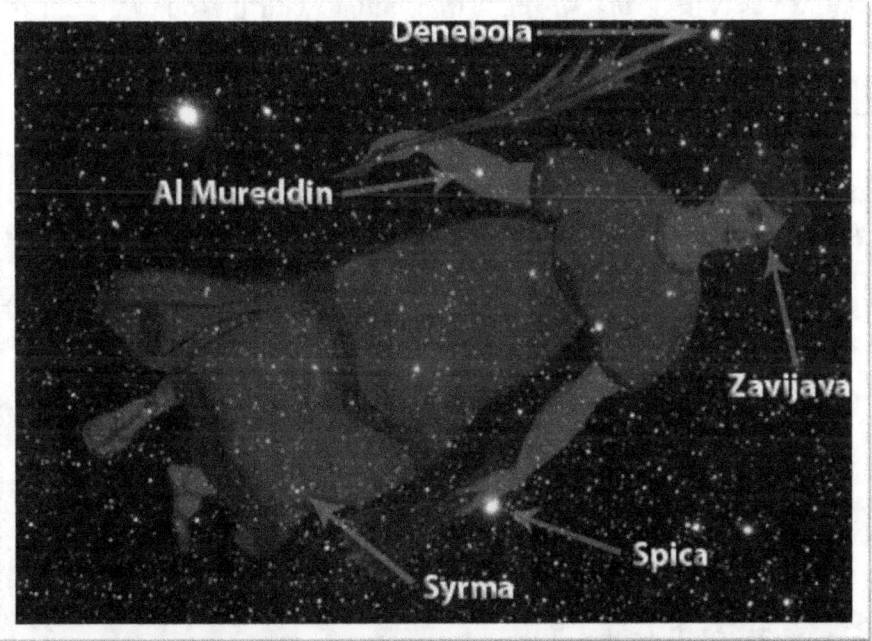

In spite of the many pagan myths taught through Greek Mythology prior to the time of Christ, the most popular was that a manifestation of Deity within the human race was soon to come on the scene. These teachings included that the supreme God would be unveiled through an immaculate conception (be born to a virgin mother). This divine child would have his life sought out by an evil being, pictured in Astrology as a serpent. This serpent would slay the Incarnate Son and in so doing fulfill the ultimate purposes for the supreme God.

These facts are a grand introduction to the first of twelve main constellations, Virgo. In the portico of *the Temple of Esneh* in Egypt there is a great astronomical chart on the ceiling which shows the entire Zodiac with all its constellations. Between the constellations of Leo and Virgo, is an engraving of a

1

Sphinx with a female's head facing Virgo with a Lion's tail pointing to Leo. When this was discovered, it helped astronomers better understand the start and finish of the ancient Zodiac.[9] Thus the first constellation to be known is Virgo, *the virgin*. Virgo, along with the other eleven main constellations, has three decans. They are Coma, Centaurus and Bootes. When put together, they present an amazing first chapter for the story of *Biblical Astrology*.

Virgo compliments the oldest prophecy in the Scripture (Genesis 3:15). One that we will see referred to over and over again throughout the course of *Biblical Astrology*. "And I will put enmity between thee and the woman, and between thy seed and her seed; it shall bruise thy head, and thou shalt bruise his heel." In this prophecy, we see three noteworthy truths:

> (1) Through the seed of the woman would come the One that would bruise the Serpent's head.
> (2) Humanity did not need to be saved from itself, but from the power of its great subjugator (Heb. 2:14; Act. 26:18).
> (3) A price (i.e. bruising; cf. Is. 53:5) was to be paid for sin's curse. This price would become known to the human race as the sacrifice of Jesus Christ at Calvary.

From *the Fall of Man*, humanity would travel down the road of time over 5,000 years before this "Seed" would finally be born.[10] Throughout this time, God sent reminders of "the Coming Seed Promise" through the prophet Isaiah (792 B.C.) and others.[11] "Therefore the Lord himself shall give you a sign; Behold, a virgin shall conceive, and bear a son, and shall call his name Immanuel" (Isaiah 7:14). "For unto us a child is born, unto us a son is given: and the government shall be upon his shoulder: and his name shall be called Wonderful, Counsellor, The mighty God, The everlasting Father, The Prince of Peace" (Isaiah 9:6).

What Virgo's Stars Teach

In Virgo's right hand she holds *a branch* and in her left hand she holds *sheaves* with *wheat seeds*. With that in mind, notice what Christ said in

---

[9] *The Real Meaning of the Zodiac*, D. James Kennedy, Ph.D., Coral Ridge Ministries, Fl., 1997.

[10] Using the Septuagint version over the Massoretic.

[11] See also Ps. 132:11; Jer. 31:22; Mic. 5:3; Dan. 9:26; etc.

John 12:23-24. "And Jesus answered them, saying, The hour is come, that the Son of man should be glorified. Verily, verily, I say unto you, Except a corn of wheat fall into the ground and die, it abideth alone: but if it die, it bringeth forth much fruit." Not only was Christ teaching Andrew and Philip the philosophy that giving your life is more profitable than hording it, but he was also teaching, in the saying, "the hour is come, that the Son of man should be glorified," that he would give himself over to death in order to bring forth much fruit. Significantly, the definite article is present in the Textus Receptus in John 12:24 anteceding "corn." Since Christ was referring to himself, he identified himself to "the" corn of wheat, which is the very thing Virgo holds in her hand.

Interestingly enough, the brightest star in Virgo is *Spica*; which comes from the Arabic word *Al Zimach,* meaning "the seed (or corn) of wheat."[12] It was this corn of wheat that would die in order to bring forth what Virgo holds in her other hand, *the Branch*. Through this Branch, the Christian Church would be grafted in (Rom. 11:17-24) and flourish; bringing much fruit into eternal life. In 519 B.C., Zechariah prophesied of the Branch, saying, "And speak unto him, saying, Thus speaketh the LORD of hosts, saying, Behold the man whose name is The BRANCH; and he shall grow up out of his place, and he shall build the temple of the LORD" (Zechariah 6:12).

The doctrine of Christ's *Virgin Birth* has been under attack from Satan from its conception because it validates Genesis 3:15's *Coming Seed Promise*. The Virgin Birth guarantees the believer that Christ's shed blood would be sinless and a means of justification for what Satan caused in the Garden of Eden. It also prompts believers to reason Christ's Sacrifice as a holy substitution for their deserved *Condemnation*. When thinking of the *Doctrine of Substitution*, one should understand that Christ stood in the believer's place, bearing his sin debt.

Although Christ's heel was bruised at Calvary, by crushing the serpent's head, he broke the serpent's power as prophesied in Genesis 3:15. Hebrews 2:14-15 says, "Forasmuch then as the children are partakers of flesh and blood, he also himself likewise took part of the same; that through death he might destroy him that had the power of death, that is,

---

[12] *Mazzaroth*, Frances Rolleston, London: Rivingtons, Waterloo Place, 1862

the devil; And deliver them who through fear of death were all their lifetime subject to bondage." From the beginning there has been a conflict between the forces of Light and Darkness.

The constellation of Virgo shows us some of the most important Christian doctrines. Virgo teaches us *the Incarnation* of deity, *Substitutionary Atonement*, *the Virgin Birth*, *the Immaculate Conception*, the *Infallibility* of the Word of God and the foundation of the Christian Church in light of Christ's death.

These great truths are speaking to us today, showing that *the Coming Seed* truly was Jesus Christ. Matthew 1:21-23 "And she shall bring forth a son, and thou shalt call his name JESUS: for he shall save his people from their sins. Now all this was done, that it might be fulfilled which was spoken of the Lord by the prophet, saying, Behold, a virgin shall be with child, and shall bring forth a son, and they shall call his name Emmanuel, which being interpreted is, God with us."

VIRGO

Score

Piano

Star Chart Music

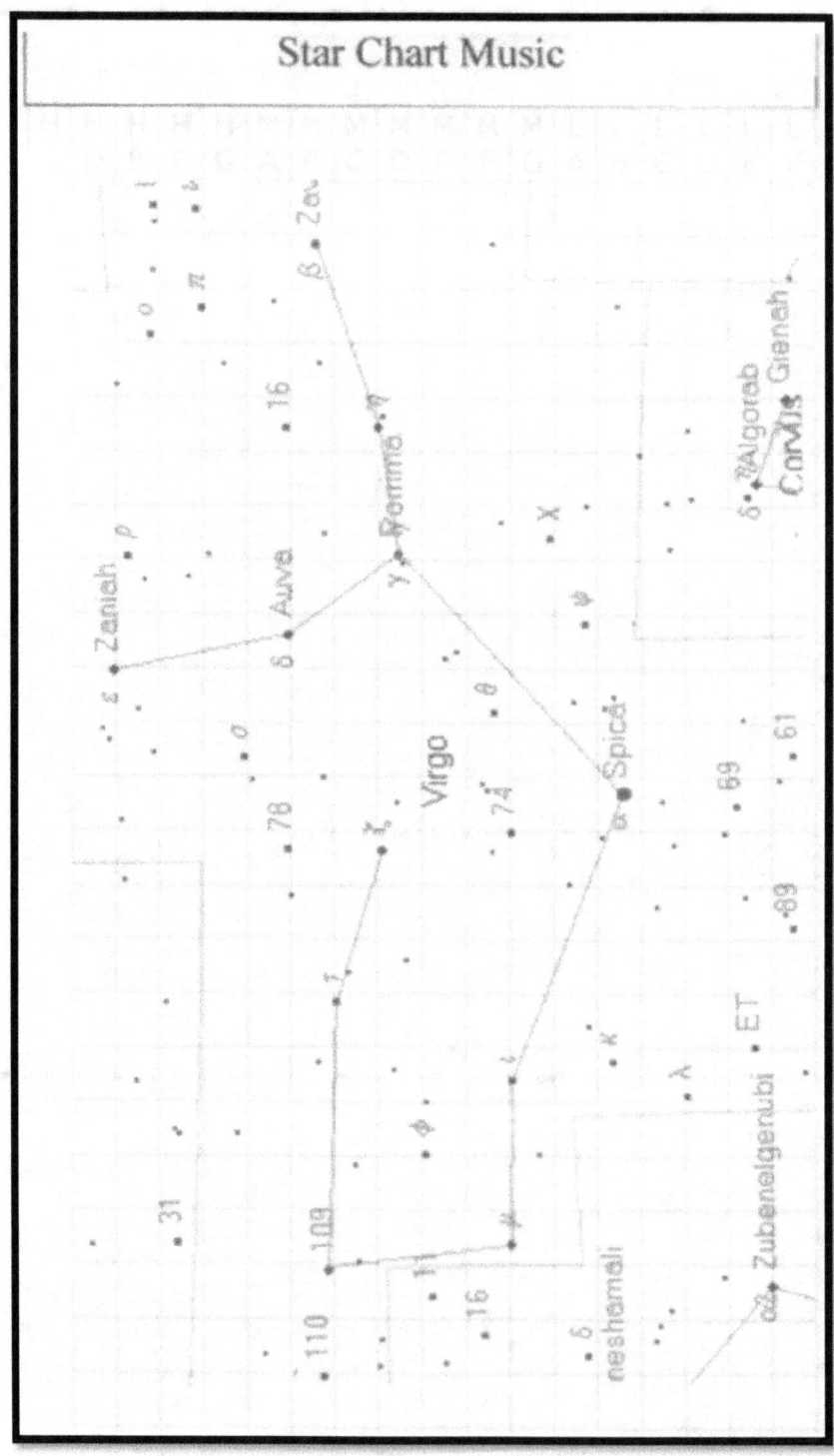

COMA "THE DESIRED"
CHAPTER 1 SECTION 2

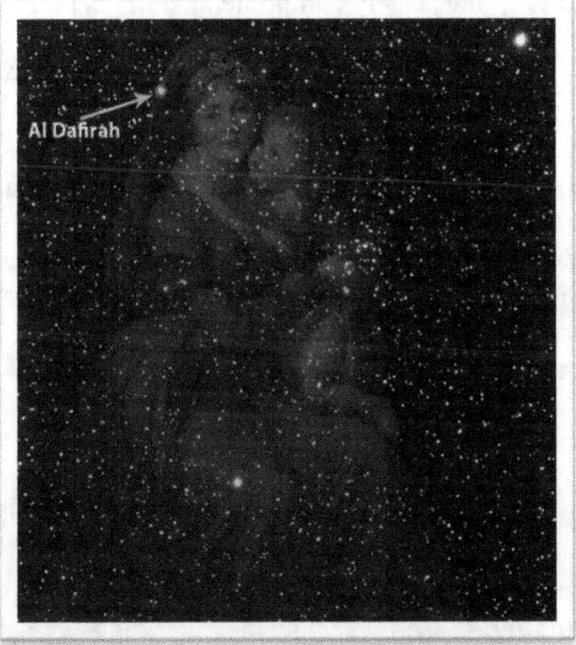

The Hebrew name for Coma means *desired*. The Greeks perverted this constellation by transliterating the word and then giving it an interpretation. Therefore, many modernists view this constellation as the wig or hair of Berenice, signifying the wife of Euegretes (Ptolemy III) who had her hair stolen.[13] The zodiac in the Temple of Denderah in Egypt (2,000 B.C.) reveals this constellation as a woman holding her child. The name ascribed to it by the ancient Egyptians was *Shes-nu,* meaning "the desired son!"

> Haggai 2:6-7 For thus saith the LORD of hosts; Yet once, it *is* a little while, and I will shake the heavens, and the earth, and the sea, and the dry *land*; And I will shake all nations, and **the desire of all nations** shall come: and I will fill this house with glory, saith the LORD of hosts.

---

[13] *Constellations of Words*, Explore the etymology and symbolism of the constellations, http://www.constellationsofwords.com/Constellations/ComaBerenices.html

The statement in Haggai "the desire of all nations" has been under historical examination ever since it was penned. Most commentaries are in debate as to what was meant. It is my guess that Haggai was referring to Coma (520 B.C.); thus, easily comprehended in his present day.

One Arabic astronomer, named *Albumazer* (805-885 A.D.) said the following concerning this constellation:

> There arises in the first Decan, as the Persians, Chaldeans, and Egyptians, and the two Hermes and Ascalius, teach a young woman whose Persian name denotes a pure virgin sitting on a throne, nourishing an infant boy having a name, by which, some nations call Ihesu, with the signification Ieza, which in Greek is called CHRISTOS.[14]

Virgo symbolized that *the Coming Branch* would be born of the seed of the woman. Coma, a decan of Virgo, tells us that that Branch would be a child. In this story, we see a stellar type of Israel as in Revelation 12:1-8.

> And there appeared a great wonder in heaven; a woman clothed with the sun, and the moon under her feet, and upon her head a crown of twelve stars: And she being with child cried, travailing in birth, and pained to be delivered. And there appeared another wonder in heaven; and behold a great red dragon, having seven heads and ten horns, and seven crowns upon his heads. And his tail drew the third part of the stars of heaven, and did cast them to the earth: and the dragon stood before the woman which was ready to be delivered, for to devour her child as soon as it was born. And she brought forth a man child, who was to rule all nations with a rod of iron: and her child was caught up unto God, and *to* his throne. And the woman fled into the wilderness, where she hath a place prepared of God, that they should feed her there a thousand two hundred *and* threescore days. And there was war in heaven: Michael and his angels fought against the dragon; and the dragon fought and his angels, And prevailed not; neither was their place found any more in heaven. (Revelation 12:1-8)

---

[14] *Albumazer* (805-885 A.D).

The woman in the heavens is royal indicating she's Israel and her child will descend through the Davidic Dynasty. The boy child is none other than this Woman's Messiah. During the times of Christ, Israel was looking for and desiring her Messiah (Mt. 2:1-7; Jn. 1:41; 4:25; 4:42; etc.), but when he came unto her, Satan—through Judaic prejudice and evil government—blinded Israel's eyes and got her to reject her Messiah (Jn. 1:11).

CENTAURUS
CHAPTER 1 SECTION 3

One of the most amazing stories pictured in the heavens is the story of Centaurus. In this recital, we see the evil and dual natured Centaur slaying Lupus "the victim." This victim is the Son of the Virgin that we discussed in the previous sections.

Principal Star Names

Al-Beze is an Arabic name of Centaur, meaning "the despised."[15] Many biblical astrologers have tried to identify Centaurus as Christ by saying that *Al-Beze* refers to the Isaiah 53:3 prophecy "he is despised and rejected of men." In addition, they add the verse in John 10:18 which says, "No man taketh it from me, but I lay it down of myself. I have power to lay it down, and I have power to take it again." Alexander Jamieson said,

> On the authority of the most accomplished Orientalist of our own times, the Arabic and Chaldaic name of this constellation is Bezeh." Now this Hebrew word Bezeh (and

---

[15]Mazzaroth, Frances Rolleston, London: Rivingtons, Waterloo Place, 1862

the Arabic Al Beze) means the despised. It is the very word used of this Divine sufferer in Isaiah 53:3, "He is DESPISED and rejected of men."[16]

Notice though, Christ was given the power to lay down his life, but not to kill it. This view has a few shortcomings. First, it makes Centaur and the Victim one in the same. Although Christ was despised by the world, Satan is the despised of the church, in that he is the arch-enemy of God's kingdom. A sharper eye will reveal that Christ **laid down** his life for his enemies to kill, but did not take his own. This is seen in Peter's sermon on Pentecost, "Him, being delivered by the determinate counsel and foreknowledge of God, ye have taken, and by wicked hands have crucified and slain" (Acts 2:23). Although the Cross brings about the righteousness of God, it also brings out the sinister evil of Satan; therefore, both forces represented here. So, the *despised one* is none other than Satan, and the Victim is none other than Jesus Christ.

Secondly, it denies the repetitious conflict seen throughout the stars nearly 10 other times. The narrative within the star's record cannot be ignored. It echoes the conflict of Satan and Christ. To view this one constellation opposite of the rest would reconstruct the interpretive method of viewing the stars.

Thirdly, in addition, the Greek name for Centaur is Cheiron, meaning "one who pierces." This identifies Centaurus as an enemy, when comparing such scriptures as Psalm 22:16, "For dogs have compassed me: the assembly of the wicked have inclosed me: they pierced my hands and my feet." Here we see the prophetic announcement, not that Christ would pierce himself, but that he would be pierced by "the wicked." After everything he endured up to this point, we still see *the Victim of sin's circumstance* laying down his life as a Lamb to the slaughter.

Toliman – a star located on the hoof of Centaur meaning "the heretofore and hereafter." Looking at the images the stars have painted, we find that the Victim has been trampled under the feet of Centaur; therefore, the star in the hoof identifies the one trampled, namely "the heretofore and hereafter." Hebrews 10:28-29 says "He that despised Moses' law died

---

[16] *Celestial Atlas*, Alexander Jamieson, 1822, https://astrologyking.com/constellation-centaurus/

11

without mercy under two or three witnesses: Of how much sorer punishment, suppose ye, shall he be thought worthy, who hath trodden under foot the Son of God, and hath counted the blood of the covenant, wherewith he was sanctified, an unholy thing, and hath done despite unto the Spirit of grace?"

Centaurus is also a picture of Israel under Satanic influence. *Toliman* (located in the hoof or heal; see Gen. 3:15) is a picture of Jesus Christ, the "Alpha and Omega, the beginning and the ending, saith the Lord, which is, and which was, and which is to come, the Almighty" (Rev. 1:8). Is there any wonder why this star is the brightest star in this constellation? This trampling that Christ endured at Calvary includes the beating that he took in humanity's place. "As many were astonied at thee; his visage was so marred more than any man, and his form more than the sons of men" (Is. 52:14). The Bible tells us here that Christ was beaten beyond recognition. His face was beaten so bad that his lips were swollen and busted, his eyes were blackened and bloodied, and his nose (which is not a bone, but cartilage) was probably broken. In addition, his body was scourged; meaning, his back was ravaged by a whip called the cat of nine tails. This kind of scourging caused the raw tissue to come out of the skin. After all this, he had enough strength to still carry his cross for a little while (Matt. 27:32).

Hadar is the star located on the other hoof of Centaur and surprisingly means "splendor, honor, and glory." It is because of the crucifixion that Paul announced, "but God forbid that I should glory, save in the cross of our Lord Jesus Christ" (Gal. 6:14a). Likewise, it is through the blood sacrifice of Christ that all of Christendom has its glory. Why are we so attracted to the suffering, beating, blood, crucifixion, and death of Christ? Only a Christian can understand these things.

We also see the sign of the Southern Cross (Crux) tilted unjustly in favor of the Centaur. This cross signifies the innocence and justness of the slain victim. Is it not odd that Christ was declared innocent and just by Pilate, the head of the Judean province in the Roman Empire? In Luke 23:4, Pilate said "I find no fault in this man" (Lk. 23:4). Furthermore, we see "When Pilate saw *that* he could prevail nothing, but that rather a tumult was made, he took water, and washed *his* hands before the multitude, saying, I am innocent of the blood of this **just** person: see ye *to it*. Then answered all the people, and said, His blood *be* on us, and on our children" (Matt. 27:24-25). Further signifying Christ's innocence, Pilate put above the head of Christ a placard which read, "JESUS OF NAZERTH KING OF THE JEWS" (Jn. 19:19). The chief priests did not like this and demanded that it be changed. These words were held in the Latin letters INRI; however, the Hebrew gave the acronym for YHVH, which transliterated is Yahveh. These four Hebrew letters are called the *Tetragrammaton* and are pronounced in English as *Yod, Hay, Vav, Hay;* matching perfectly with John 19:19. The sign in *Aramaic Hebrew* would have been pronounced in English as Yeshua (Jesus) – Hanatzree (Nazereth) – Vemelek (King) – Hayehoodeem (Jews). This acronym pronounced the sacred and incommunicable name of God. By placing it above the head of Christ, Pilate was affirming Christ's identity as Jehovah.[17] With the story of Christ in mind, we have a magnificent disclosure of the Victim as being Christ himself. Surely, Jesus was "the Lamb slain from the foundation of the world" (Rev. 13:8).

It is no surprise that nearly twenty times in the astrological record we can see this very story pictured. Centaurus symbolized, can be none other than Israel under the satanic influence of the Devil (Matt. 26:57-59). Satan always needs a system, religion, nation or body to work through. It is worth noting in the story of Christ, that it was the conservative group called "the Sanhedrin" that persecuted Christ.

Concerning Satan's Multiple Natures

In the dual nature of Centaurus being half horse and half man, we can better understand the subtle ability of Satan to morph into or within "an angel of light" (II Corinth. 11:14). The Bible infers that Satan was at one time an arch angel of the most magnificence. The Bible says, "Thou sealest up the sum, full of wisdom, and perfect in beauty" (Ez. 2812). In his primitive life, Satan was an angel of great power. "Thou *art* the anointed cherub that covereth; and I have set thee *so*: thou wast upon the holy mountain of God; thou hast walked up and down in the midst of the stones of fire" (Ez. 28:14).

---

[17] The more accurate pronunciation of Jehovah is Yahveh, since Hebrew has no "j" sounding letter as seen in Jehovah, or "w" sounding letter as in Yahweh.

13

Concerning "the Victim"

It is significant to note that the Victim is pictured upon his back as being trampled by the front hooves of Centaurus. Ulugh Beigh said that the Victim was anciently called *Sura*, meaning *a sheep or lamb*.[18] One may ask, why did Christ endure the crucifixion when he could have prayed to his Father and received more than twelve legions of angels (Matt. 26:53)? The first reason for Christ's suffering was that the Scriptures must be fulfilled (Matt. 26:54).

Concerning Crux "the Southern Cross"

There were many prophecies concerning Christ mentioned in the Old Testament, but there is one in particular that is highlighted in *the Southern Cross*, also called *Crux*. Unlike most crosses which are level, we see in Crux an unlevel cross, signifying an unjust deed. *Aben Ezra* tells its Hebrew name as *Adom*, which means *"cutting off."* All of these acts were simultaneous: a victim slain, by ungodly hands, unjustly, by Satan and his band. This was an amazing fulfillment of Daniel 9:26 "And after threescore and two weeks shall Messiah be cut off, but not for himself: and the people of the prince that shall come shall destroy the city and the sanctuary; and the end thereof *shall be* with a flood, and unto the end of the war desolations are determined." Shortly after the Messiah was cut off, the city and sanctuary were destroyed under Titus Vespasian in A.D. 70. This concluded the Old Testament Prophecy of Daniel 9:26 and the stellar Lamb that was slain from the foundation of the world. According to Daniel 9:25, these seventy weeks were to begin "from the going forth of the commandment to restore and to build Jerusalem." This took place in Ezra chapters 7-8 (457 B.C.). (1) The "seven weeks" (Dan. 9:25) or *forty nine years*, is the period which saw the restoration of Jerusalem recorded in the books of Ezra and Nehemiah. (2) The "three score and two weeks," or *four hundred and thirty four years,* is the period that leads to the time of Christ. (3) The "one week," or *seven years of tribulation,* is a period of time that is yet future, which will usher in the Second Coming of Christ.

This prophecy has its boundaries typed out in Daniel 9:24. "Seventy weeks are determined upon thy people and upon thy holy city." This prophecy does not primarily concern the Church, but it does through a

---

[18] *Ulegh Beg's Catalogue of Stars*, Revised from Persian Manuscripts Existing in Great Britain, with a Vocabulary of Persian and Arabic Words, By Edward Ball Knobel, The Carnegie Institution of Washington, 1917.

secondary viewpoint. The church is not mentioned in the *Tribulation Period* because we were *cut off* with Christ and are, therefore, partakers of his suffering (2nd Cor. 1:7; Heb. 3:14; 1st Pet. 4:13). Needless to say, there still remains a week of years left. This is referred to by theologians and Christians as the coming *Tribulation Period.*

What a powerful inter-correlation of prophecy and primeval astrology. At the command of Artaxerxes, around 457 B.C., this prophecy of the 69 weeks (483 years) began. Most scholars believe that Christ was born in 4 B.C., namely "in the days of Herod the Great" (Matt. 2:1). Luke 3:23 says that Jesus was *around thirty years old* at the time when he began his three years of earthly ministry. Therefore, Christ's complete life was between 33 and 36 years. By subtracting 453 years from 457 B.C., we land at 4 B.C. Subtracting the same number from our 483 years to the cutting off of the Messiah, we have a remainder of 30 years, which is close enough to an exact and precise fulfillment. From 457 B.C. to the birth of Christ (4 B.C.) is 453 years. Again, subtracting 453 from our 483 (69 x 7) years, we have a remainder of 30 years till the cutting off. Some have shown the 69 weeks' prophetic fulfillment when Christ started his ministry when he read Isaiah 61 in Luke 4:17-21 and others have shown its fulfillment at Calvary.

The Bible informs us that the second reason for Christ's pain and agony was God's love. "For God so loved the world, that he gave his only begotten Son, that whosoever believeth in him should not perish, but have everlasting life" (John 3:16). Christ understood the subjugation that took place within the Fall of Man and therefore wanted to redeem the world from this inherent sin and dominion of Satan (Acts 26:18) by making his life a ransom. The whole story of Christ's suffering is one that demonstrates the love of God. Although many sinners have come to repentance, only those that have come to terms with God's love have made it to salvation. No one in the world could have been saved had Christ not gone to Calvary.

15

At Calvary

Years I spent in vanity and pride,
Caring not my Lord was crucified,
Knowing not it was for me He died on Calvary.

Mercy there was great, and grace was free;
Pardon there was multiplied to me;
There my burdened soul found liberty at Calvary.

By God's Word at last my sin I learned;
Then I trembled at the law I'd spurned,
Till my guilty soul imploring turned to Calvary.

Mercy there was great, and grace was free;
Pardon there was multiplied to me;
There my burdened soul found liberty at Calvary.

Now I've given to Jesus everything,
Now I gladly own Him as my King,
Now my raptured soul can only sing of Calvary!

Mercy there was great, and grace was free;
Pardon there was multiplied to me;
There my burdened soul found liberty at Calvary.

Oh, the love that drew salvation's plan!
Oh, the grace that brought it down to man!
Oh, the mighty gulf that God did span at Calvary!

Mercy there was great, and grace was free;
Pardon there was multiplied to me;
There my burdened soul found liberty at Calvary.

William Newell, 1895

Bootes "The Coming One"
Chapter 1 Section 4

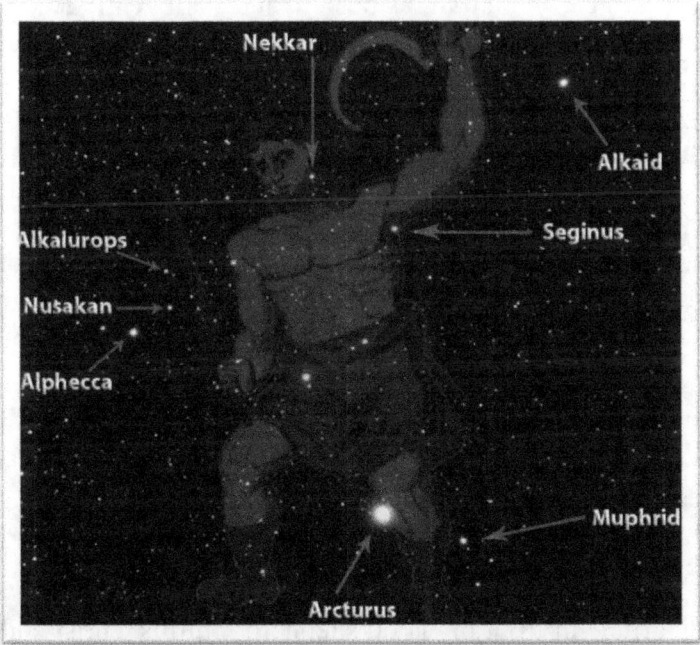

Bootes is the third decan in Virgo which expresses the gospel story in a marvelous way. So far, the heavenly images have shown us Christ's entrance through the nation of Israel, his deity, Satan's plot, and the Crucifixion. *Bootes* is a Hebrew derivative which means *"the coming one."*[19] Bootes was the resurrected Victim. We recognize this story in the doctrine of *the Second Coming*. The principal star in this decan is **Arcturus** which similarly means *"He Cometh."* It is the ancient name for Bootes and is mentioned in Job. In Job 9:9, Job said, "Which maketh **Arcturus**, Orion, and Pleiades, and the chambers of the south." In Job 38:32, God said "Canst thou bring forth Mazzaroth in his season? or canst thou guide **Arcturus** with his sons?" According to ancient astronomy, Bootes (Arcturus) was the guardian, watchman, and keeper of Arktos (the sheep fold or sons) which now are the adjoining constellations known as the greater and lesser bears or the big dipper (Ursa Major) and the little dipper (Ursa Minor) picturing beautifully God's nation Israel, and the Church.

---

[19] Mazzaroth, Frances Rolleston, London: Rivingtons, Waterloo Place, 1862

One of the most expressive figures under which Christ is pictured in the Old Testament is that of an Oriental Shepherd. Isaiah announced Christ as One who "shall feed his flock like a shepherd: he shall gather the lambs with his arm, and carry them in his bosom, and shall gently lead those that are with young" (Isaiah 40:1). Christ said of himself and his sheep in John 10:11-15, 27-28,

> I am the good shepherd: the good shepherd giveth his life for the sheep. But he that is an hireling, and not the shepherd, whose own the sheep are not, seeth the wolf coming, and leaveth the sheep, and fleeth: and the wolf catcheth them, and scattereth the sheep. The hireling fleeth, because he is an hireling, and careth not for the sheep. I am the good shepherd, and know my sheep, and am known of mine. As the Father knoweth me, even so know I the Father: and I lay down my life for the sheep... My sheep hear my voice, and I know them, and they follow me: And I give unto them eternal life; and they shall never perish, neither shall any man pluck them out of my hand.

The definite article in front of "good shepherd" fulfills Christ's identity as the Shepherd of Old Testament prophecy and possibly the heavenly Bootes. Here, we can conclude that our Shepherd didn't come to make his sheep sacrifice their lives for him, but that he would give his life for them. This is pictured in the head of Bootes by another principal star called *Nekkar*, which means "the pierced." In the scripture reference, the definite article is also present in front of *wolf,* which identifies him as the arch enemy of the Shepherd and the sheep. Notice in Christ's words that he does not permit the wolf's presence, but promises redemption and deliverance from Satan.

The Staff and Sickle

The shepherd uses a staff called a *shepherd's crook* to guide, protect and love his sheep. *Al-Katurops* is a star by the right arm of Bootes, which means "the branch, rod or staff," perfectly representing the shepherd's crook.

The shepherd's crook is a good representation of God's Word. Through his staff, Bootes is reminding us to stay close to the Word of God for guidance and protection. It may be old, but it's still "quick (alive), and powerful, and sharper than any two-edged sword" (Hebrews 4:12). This staff shows us God's love, guides us through the valleys of life and is the Christian's safeguard from the Wolf.

In addition to the staff, Bootes also bares a sickle, which is identified by a star named *Muphride*, meaning *he who separates*. *Al-Katurops* and *Muphride* show the multi-tasking of the Shepherd. Not only is he the great Shepherd of the sheep, but also the great Reaper of the harvest. His harvest is not one of food, but of souls. In Christ's only prayer request, he admonished his disciples to "pray ye therefore the Lord of the harvest, that he will send forth labourers into his harvest. Then saith he unto his disciples, the harvest truly is plenteous, but the labourers are few" (Matthew 9:37-38). The Christian's duty to go and share the love of Christ has been the ultimate challenge in the Church Age. As easy as it is to get sidetracked, we see in this constellation a reminder by the Great Shepherd to stay busy at the Great Commission no matter what, knowing that our labor is not in vain. Through his sickle, Bootes is telling us to hold out till the end. Although we may be in the latter days, there is still a harvest to reap. In our reaping, let's be admonished of Christ's command to not root up the tares, lest we root up the wheat also.

> Another parable put he forth unto them, saying, The kingdom of heaven is likened unto a man which sowed good seed in his field: But while men slept, his enemy came and sowed tares among the wheat, and went his way. But when the blade was sprung up, and brought forth fruit, then appeared the tares also. So the servants of the householder came and said unto him, Sir, didst not thou sow good seed in thy field? from whence then hath it tares? He said unto them, An enemy hath done this. The servants said unto him, Wilt thou then that we go and gather them up? But he said, Nay; lest while ye gather up the tares, ye root up also the wheat with them. Let both grow together until the harvest: and in the time of harvest I will say to the reapers, Gather ye together first the tares, and bind them in bundles to burn them: but gather the wheat into my barn. (Matt. 13:24-30)

My Shepherd

He fought with and defeated the wolf on Calvary's rugged cross;
He came to seek, find, and save his sheep that he knew were lost.

He guides them into shade, during summer's heat,
And provides for them their needed drink and meat.
He comforts them in the winter's cold;
He's been gentle, keeping them in his fold.

He's cried o'er all that e'er went away,
Hanging on to them in prayer, by night and by day.
He's been a watchman all around the clock;
He's never disowned one from among his great flock.

Soon, may that flock begin to worship and sing,
And let the praises for his love's sake ring;
What a Great Shepherd is Jesus Christ our King!

Our Good Shepherd is presently at the right hand of the Father and "he ever liveth to make intercession" on our behalf (Hebrews 7:25). That means He is praying for us. Christ is praying for deliverance from the power of the Wolf through faith in Jesus Christ as our personal Savior, for our safe return to the flock when we get sidetracked and drawn away from God and for us to be yielded that he may accomplish his purpose in our lives.

Hebrews 13:20-21 "Now the God of peace, that brought again from the dead our Lord Jesus, that great shepherd of the sheep, through the blood of the everlasting covenant, Make you perfect in every good work to do his will, working in you that which is well pleasing in his sight, through Jesus Christ; to whom be glory for ever and ever. Amen."

Win the Lost at any Cost

As we look around us,
All the fields are white,
Ripened unto harvest,
Yet so quickly comes the night.
Christians must get busy,
There's is work to do.
Here's an urgent task awaiting you.

Souls are crying, men are dying,
Won't you lead them to the cross.
Go and find them, help to win them
Win the lost at any cost.

Check your fold, my brother,
Are all your children in.
Are there some still straying in the
Blackened fields of sin.
You must go and win them,
Go without delay,
Soon the trump of God shall end the day.

Go out and win, rescue from sin,
Day's almost done, lo sinks the sun.
Souls are crying, men are dying,
Win the lost at any cost.

Ellis / Mauldin

BOOTES

Daniel G. McCrillis
Daniel G. McCrillis

Score

Piano

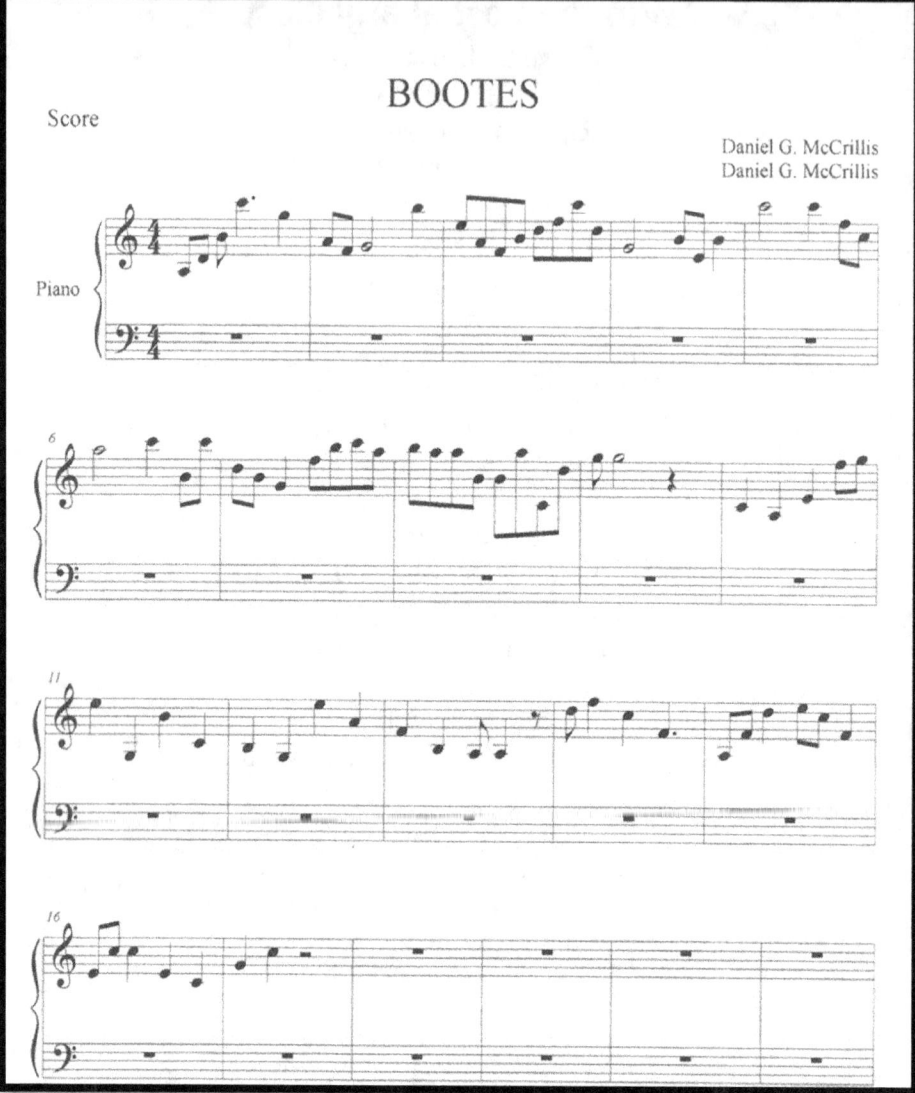

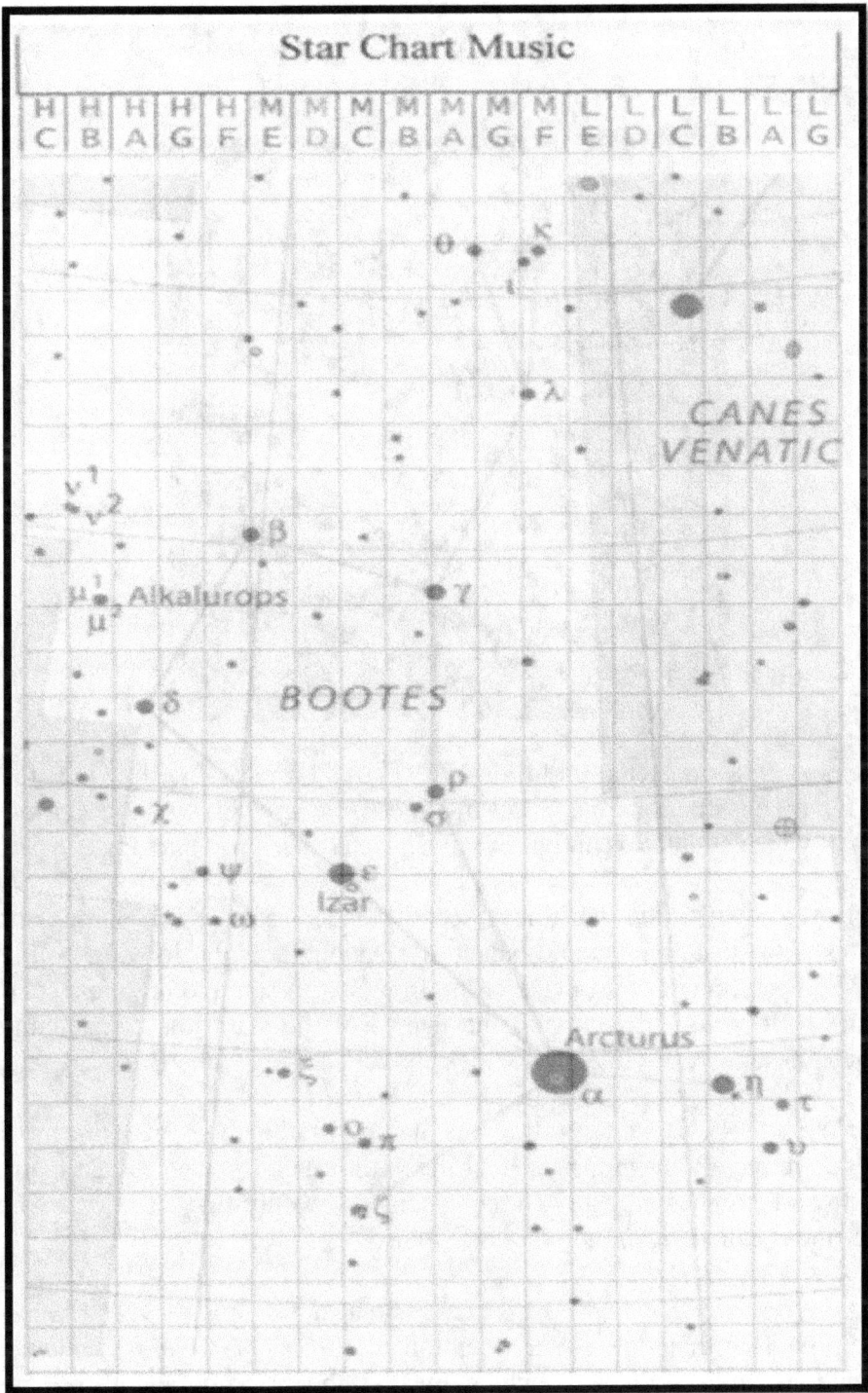

LIBRA "HEAVENLY JUSTICE"
CHAPTER 2 SECTION 1

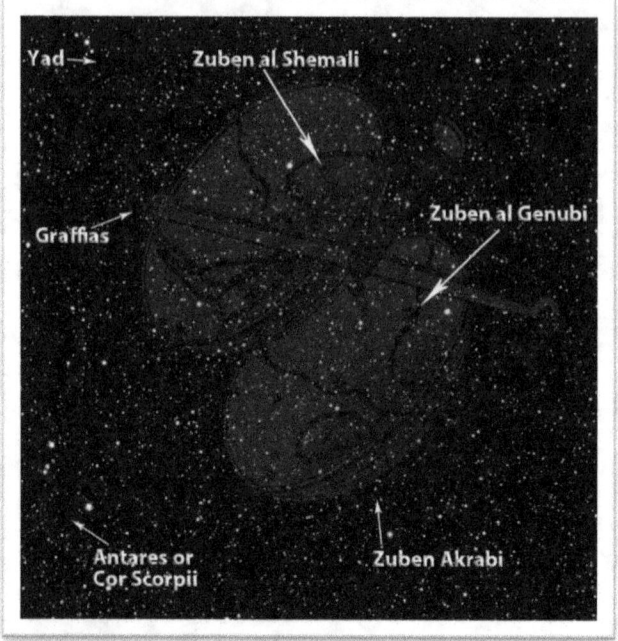

The next great constellation following Virgo is Libra. Libra, like the other 11 main constellations has 3 decans, *the Cross*, *the Victim*, and *Corona*, two of which (the Cross and the Victim), we covered in Centaurus.

Libra is known to the world as a set of balance scales. The purpose for scales is for just weights and balances. Undoubtedly, by looking up into the night sky, one cannot make a line from star to star and outline a set of scales. However, by studying the names of the stars this constellation proclaims a powerful image of scales and tells a story. The Bible says that God "telleth the number of the stars; he calleth them all by their names" (Psalms 147:4). By interpreting the names of the three major stars in Libra, we unveil an important doctrine within the Gospel message that is worthy of emphasizing to Christians in the twenty-first century. This doctrine is called *Justification of the Believer*.

"THE PRICE DEFICIENT"

The first star that draws our attention is *Zuben al Genubi*. *Zuben al Genubi* is an Arabic name which means "the price deficient" or "not

24

enough."[20] This star teaches man of his sinful state and condition. No doubt everyone realizes that there is sin in the world. This sin was inherited by the human race through Adam and Eve's transgression. The Bible says in Romans 3:23 "all have sinned and come short of the glory of God." "Come short" indicates that we are as Belshazzar and have been "weighed in the balances, and art found wanting" (Daniel 5:27). Romans 3:23 is telling us that mankind is missing the mark, deficient, and unjust before a holy God. Nothing we can do can take away the wrongs that we've committed. Thus, we see *the Depravity of Man*, in that he is a sinner by nature and unable to pay his own sin debt. The *Depravity of Man* doctrine falls under *the Fall of Man* and leads us to the next star in Libra which is *Zuben Akrabi*.

"THE PRICE OF THE CONFLICT"

Zuben Akrabi is an Arabic name which means "the price of the conflict."[21] God, knowing man was unable to justify himself, sent his Son to suffer and die for the sin of humanity. This is called the doctrine of *Substitution* or *the Vicarious Death*. The Bible says in John 3:16, "For God so loved the world that he gave his only begotten son that whosoever believeth in him should not perish but have everlasting life." This was the price of the conflict (Zuben Akrabi). The price for redeeming the fallen race cost God his only begotten Son. This was the only price that God deemed justifiable to redeem man's sinful state. Because of Christ's Deity his blood was sinless. Sinless blood became the redemptive qualifier for the human race; thus, the only just payment for man's sin in the eyes of God. Paul, writing to the church at Corinth, said, "For he (God) hath made him (Jesus) to be sin for us, who knew no sin; that we might be made the righteousness of God in him" (II Corinth. 5:21).

"THE PRICE WHICH COVERS"

The third star demanding our attention is *Zuben al Shemali*. *Zuben al Shemali* is an Arabic name which being interpreted means "the price which covers or atones."[22] When man realizes that no matter how much good he does it will never undo the bad he has already done, then he can turn to the power of Christ's blood to have his sins atoned. With that in mind, we should realize that it would be unjust for God to give people

---

[20] Mazzaroth, by Francis Rolleston, 18
[21] Ibid.
[22] Ibid.

eternal life who reject Jesus Christ. This is why the teaching of *Works Salvation* is destructive to God's plan of salvation. Think about this question. If man could ever do anything to merit justification with God, then why did God have to send his Son to die on the Cross in order to justify our sin? The only way man can be justified is through realizing his depraved state, turning to the finished work of Christ, and receiving that work as his hope for salvation.

Christians should do good works because they are saved, not in order to be saved. Our justification took place in Heaven where God dwells. The Bible says that Christ, after his crucifixion, became our "high priest of good things to come, by a greater and more perfect tabernacle, not made with hands, that is to say, not of this building; Neither by the blood of goats and calves, but by his own blood he entered in once into the holy place, having obtained eternal redemption *for us*" (Heb. 9:11-12). When Jesus told Nicodemus in John 3:3 "Verily, verily, I say unto thee, Except a man be born again, he cannot see the kingdom of God," he was speaking of a new birth in Heaven. The words "born again" indicate a regeneration from above. We were born down below in the flesh, but at salvation we were born from above (άνωθεν). The matter of *Heavenly Justice* ventures into many other doctrines such as salvation by grace through faith, eternal security, substitution, imputation and atonement to name a few.

In Matthew 7:21-23, Christ rebuked the works salvation gospel saying, "Many will say to me in that day, Lord, Lord, have we not prophesied in thy name? and in thy name have cast out devils? and in thy name done many wonderful works? And then will I profess unto them, I never knew you: depart from me, ye that work iniquity." While many try to work their way to Heaven, the stars in Libra are a testament to **the Finished Work of Christ.** Again, the only balanced weight that will satisfy God's justice is the blood of Jesus Christ being applied to our account. "For if the blood of bulls and of goats, and the ashes of an heifer sprinkling the unclean, sanctifieth to the purifying of the flesh: How much more shall the blood of Christ, who through the eternal Spirit offered himself without spot to God, purge your conscience from dead works to serve the living God?" (Heb. 9:13-14). The song writer said it perfect when he penned,

"O precious is the flow, that makes me white as snow,
No other fount I know, nothing but the blood of Jesus!"
(Robert Lowry)

Since God made a way to balance the scales, it would be wrong for man to take away or add to His plan. Likewise, it would be unjust for God to allow man to go into a state of non-existence after his physical death because man's wrongs would never come to justice. If non-existence was possible, then life would be unfair and God unjust; however, because of *the Conflict*, life is fair. No matter how imbalanced life seems to be down here, it all balances out with Christ and the new birth.

"Would you be free from the burden of sin?
There's power in the blood, power in the blood;
Would you o're evil a victory win?
There's wonderful power in the blood.

There is power, power, wonder working power
In the blood, of the Lamb;
There is power, power, wonder working power
In the precious blood of the Lamb!

Would you be free from your passion and pride?
There's power in the blood, power in the blood;
Come for a cleansing to Calvary's tide—
There's wonderful power in the blood.

There is power, power, wonder working power
In the blood, of the Lamb;
There is power, power, wonder working power
In the precious blood of the Lamb!

Would you be whiter, much whiter than snow?
There's power in the blood, power in the blood;
Sin stains are lost in its life giving flow—
There's wonderful power in the blood.

There is power, power, wonder working power
In the blood, of the Lamb;
There is power, power, wonder working power
In the precious blood of the Lamb!

By Lewis E. Jones

LIBRA

Daniel G. McCrillis
Daniel G. McCrillis

Score

Piano

Star Chart Music

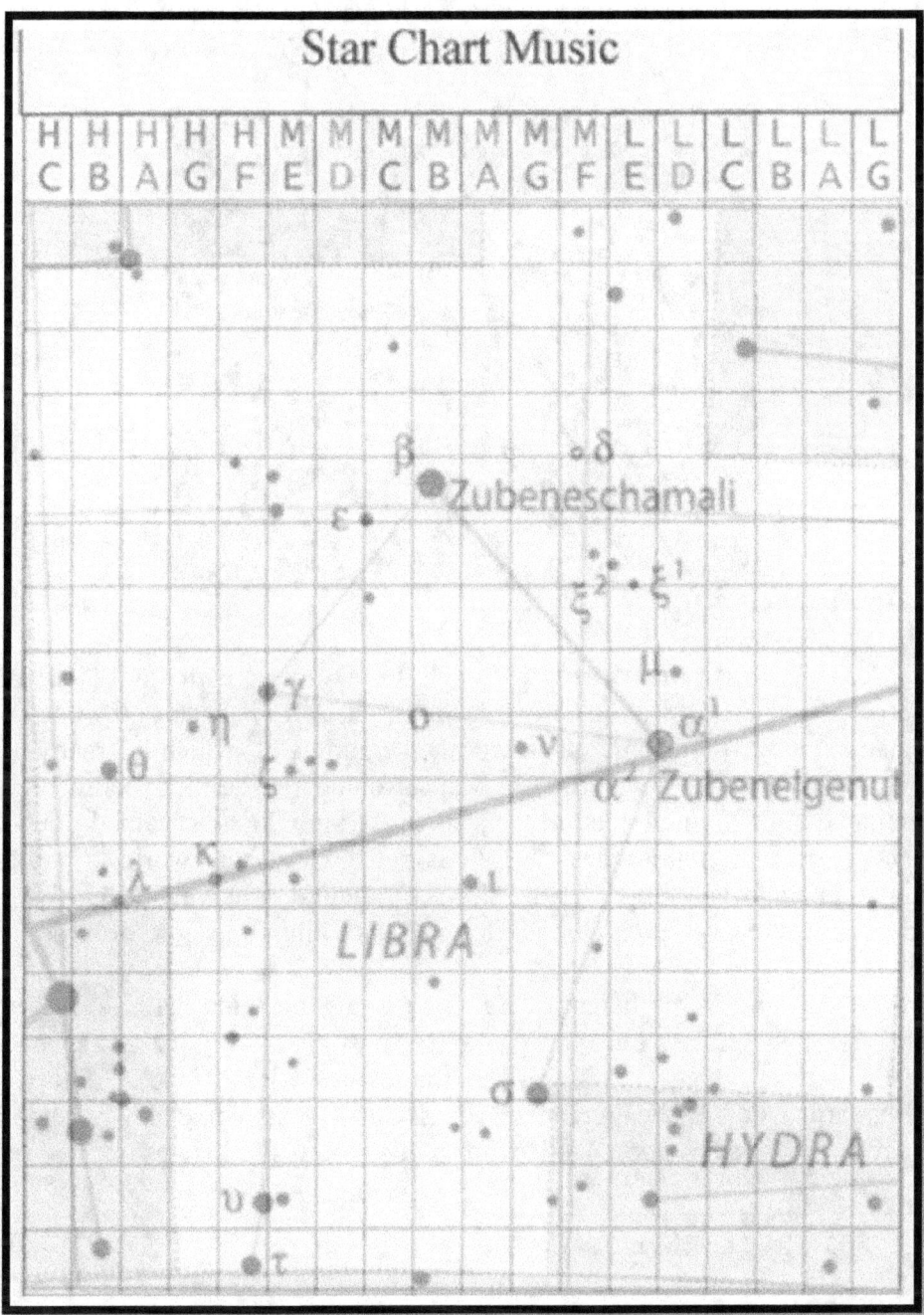

CORONA "THE NORTHERN CROWN"
CHAPTER 2 SECTION 2

Al Phecca

In Corona, we see the serpent's struggle to take the crown from God, as Serpens reaches toward the Crown (see star chart at the back of the book). Satan led a rebellion sometime after "the beginning",[23] which led many of the angels away from God. He was also the mastermind behind the subjugation of the new race in Genesis 3, during which he gained power over death and hell (hades).[24] He is the god of this world,[25] has a kingdom[26] and is a king.[27] He's still not satisfied. He wants everything that belongs to God, even God's title as King of kings and Lord of lords.[28]

In the end, and in the height of Satan's power, the saints will get weary and ask "how long, O Lord, holy and true, dost thou not judge and avenge our blood on them that dwell on the earth?" (Rev. 6:10). Shortly afterward, God, in the Person of Jesus Christ, will come down out of heaven with a sharp two-edged sword in his mouth and his vesture will be

---

[23] 1st Jn. 3:8; Rev. 12:9; Matt. 25:41.
[24] Heb. 2:14.
[25] 2nd Cor.4:4.
[26] Matt. 12:26; Lk. 11:18.
[27] Rev. 9:11.
[28] Is. 14:13-14; 2nd Thess. 2:4.

dipped in blood. At this time, he will put down all the diabolical power in *the Battle of Armageddon* and bind Satan for 1,000 years. During *the Millennial Reign*, the saints' prayer will be answered as Jesus Christ will reign "King of kings." After the 1,000 years have ended, Satan will be loosed out of his prison to pursue the crown once more, in what theologians call "*the Final Rebellion.*"[29]

PRINCIPAL STAR

The Arabic name of Corona is *Al Iclil,* meaning "ornament or jewel." The principal star in Corona is *Al Phecca,* meaning "the shining."[30] With the struggle between Satan and God over the crown, we see a visual image of the battle for the souls of men, the struggle between light and darkness, and the conflict between righteousness and evil. This crown is referring to the people from the dispensations of Law and Grace.

Exodus 19:5 "Now therefore, if ye will obey my voice indeed, and keep my covenant, then ye shall be a peculiar treasure unto me above all people: for all the earth *is* mine." Isaiah 62:3 "Thou shalt also be a crown of glory in the hand of the LORD, and a royal diadem in the hand of thy God." This is referring to Israel.

1st Peter 2:9 "But ye *are* a chosen generation, a royal priesthood, an holy nation, a peculiar people; that ye should shew forth the praises of him who hath called you out of darkness into his marvellous light." 1st Thessalonians 2:19 "For what *is* our hope, or joy, or crown of rejoicing? *Are* not even ye in the presence of our Lord Jesus Christ at his coming?" This is referring to the New Testament Church.

---

[29] *The Final Rebellion* is Satan's last attempt to overthrow the forces of righteousness. At this time he gathers the nations to battle with the Lord. In his defeat he is cast into the lake of fire and brimstone. Revelation 20:10 "And the devil that deceived them was cast into the lake of fire and brimstone, where the beast and the false prophet *are*, and shall be tormented day and night for ever and ever."

[30] Mazzaroth, Frances Rolleston, London: Rivingtons, Waterloo Place, 1862

31

SCORPIO "THE STING OF DEATH"
CHAPTER 3 SECTION 1

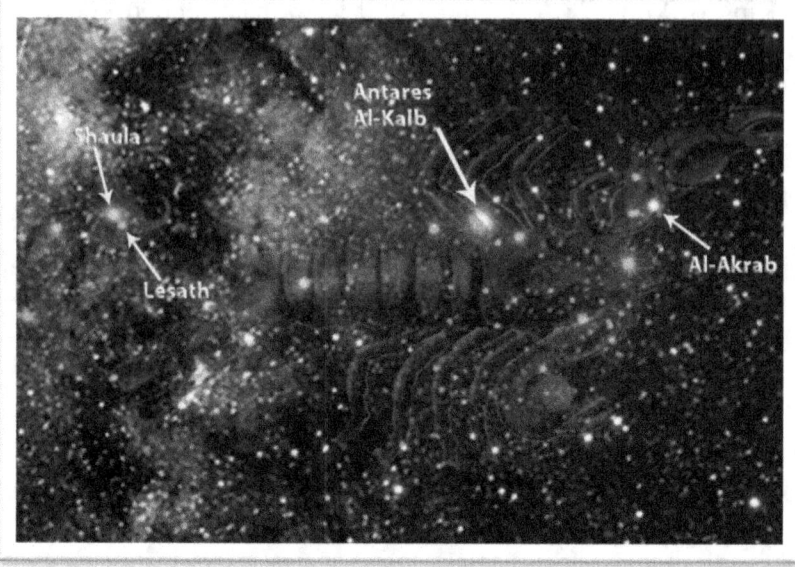

Scorpio is the third constellation in the heavens and precisely pictures Satan, the arch-enemy of God. As we can see in the image derived from the star names that Scorpio pictures a giant scorpion,[31] whose stinger is in the process of striking Ophiuchus' foot (a picture and type of Christ); Ophiuchus' other foot is simultaneously going for Scorpios' head. This fits with the Genesis 3:15 *proto-evangelism*, "And I will put enmity between thee and the woman, and between thy seed and her seed; it shall bruise thy head, and thou shalt bruise his heel." Scorpio, like the other 11 main constellations, has three decans; *Serpens*, *Ophiuchus*, and *Hercules;* all of which display an interesting part of the subject of *Biblical Astrology*.

PRINCIPAL STARS[32]

There is much insight given in the names of the stars of this giant scorpion. The Bible says that God, "telleth the number of the stars; he calleth them all by their names" (Ps. 147:4; Is. 40:26). By merely looking at the names of these stars, we have a heavenly image that contains a vital part of the gospel.

---

[31] One of the most feared creatures on the planet is the scorpion. The most poisonous scorpion carries the nickname "death stalker."

[32] E.W. Bullinger, The Witness of the Stars, Kregel Classics, 1893 reprint.

- *Al-Kalb,* is an Arabic name that means "the cleaving or conflict."
- *Antares* (located in the heart of Scorpio) is an ancient Arabic name that is thought by many to mean "the wounding." It is in Satan's heart to wound people, indicating that he is the *death stalker*, "seeking whom he may devour." In the Garden of Eden, he gained "power over death" (Heb. 2:14), and thus, has "the sting of death" (1st Cor. 15:56) in his tail. It is in his tail that he drew a third of the angels into his rebellion against God (Rev. 12:4). In the Garden, the sting of death hit the human race, causing eternal separation from God (Gen. 2:17). This was accomplished when Satan perverted God's Word (Gen 3:4) and deceived Eve to eat of the tree of the knowledge of good and evil (1st Tim. 2:14).
- *Lesath* is a bright star in the tail of Scorpio and means "the perverse" in Hebrew. This star signifies how Satan delivers his sting. Through perverseness and deception Satan works really hard at getting the human race to reason as he does. He is the great corrupter of all the good that God is and tries to do. He is responsible for perverting the human race, and the angelic.
- *Al-Akrab* is the Arabic name that means "wounding him that cometh." This is a star in the clincher of Scorpio. As the scorpion likes to grab his prey before he slams down his sting of death; likewise, Satan seeks to set up his prey to assure him that the odds are in his favor. It was in this clincher that Eve gave ear to the serpent in the Garden. Satan, in his fair speech, set her up for his inevitable sting of death. Satan set up Judas to betray Christ. He also set the atmosphere of Israel when they cried "Let him be crucified, his blood be on us and on our children" (Matt. 27:21-25). Satan was at the heart of the diabolical system behind the Herods of the Roman Empire. This system was one that would appease the religious Sanhedrin and condemn the righteous. After years of *setting the stage*, the "death stalker" was on the move; and through a series of orchestrated events, he caught his Prey (Jesus Christ) into his powerful clincher to deliver his notorious "sting of death."
- *Shaula* is a word that means "the stinger." Satan has plunged his stinger into millions of people. Although Satan perverted Judas, Herod, and the children of Israel, he never perverted Christ. Knowing the Scriptures, Christ allowed himself to be led like a lamb to the slaughter, and on that notorious day on Golgotha's Hill, Satan plunged his stinger deep into the incarnate God. At this

33

time, the infinite suffering for sin was paid for in full. Because of Christ's sinless condition, when he received his mortal wound, he was able to offer an **anti-venom** for all those who would fall victim to Satan's "sting of death." From here out, the gospel plan would be complete. It would be capable of redeeming fallen man, forgive, remit, and cleanse from sin, prove to the angels of God's great love, and finally conquer the enemy—fulfilling the Genesis 3:15's proto-evangelism. Hebrews 2:14-15 says, "Forasmuch then as the children are partakers of flesh and blood, he also himself likewise took part of the same; that through death he might destroy him that had the power of death, that is, the devil; And deliver them who through fear of death were all their lifetime subject to bondage."

Heavenly Portraits

Throughout the Scripture, we find that the Bible writers used the precise imagery of the signs of the Zodiac and its decans. The Bible says in Revelation 9:10-11, "And they had tails like unto scorpions, and there were stings in their tails: and their power *was* to hurt men five months. And they had a king over them, *which is* the angel of the bottomless pit, whose name in the Hebrew tongue *is* Abaddon, but in the Greek tongue hath *his* name Apollyon." An interesting feature of Satan is here, in that he is king of the scorpion-like creatures mentioned in Revelation.

One day, Jesus' disciples came back to Jesus and reported their successes saying, "even the devils are subject unto us through thy name" (Luke 10:17). In response to this, Jesus said, "I beheld Satan as lightning fall from heaven. Behold, I give unto you power to tread on serpents and scorpions, and over all the power of the enemy: and nothing shall by any means hurt you. Notwithstanding in this rejoice not, that the spirits are subject unto you; but rather rejoice, because your names are written in heaven" (Luke 10:18-20). It is through Christ that we have the power over the enemy and our names written in Heaven. Christ paid the infinite price when he was "made sin for us, who knew no sin; that we might be made the righteousness of God in him" (2nd Cor. 5:21). The only way man can have his name written in Heaven is if he has come through Christ's death, burial and resurrection. Christ is the door to Heaven. Belief and faith in his Word are the keys that unlock the door. Man is sinful, but Christ is sinless. Man is unjust, but Christ is just. Man is lost, but with Christ, he

34

is found. Man is born dead, but in Christ, he is born again (from above) and quickened (made alive). It is through the anti-venom that we are saved. Unlock the door through belief, and walk through by faith and receive everlasting life.

Are you Washed in the Blood?

Have you been to Jesus for the cleansing power?
Are you washed in the blood of the Lamb?
Are you fully trusting in His grace this hour?
Are you washed in the blood of the Lamb?

Chorus: Are you washed in the blood,
In the soul cleansing blood of the Lamb?
Are your garments spotless? Are they white as snow?
Are you washed in the blood of the Lamb?

Are you walking daily by the Savior's side?
Are you washed in the blood of the Lamb?
Do you rest each moment in the Crucified?
Are you washed in the blood of the Lamb?
Chorus

When the Bridegroom cometh will your robes be white?
Are you washed in the blood of the Lamb?
Will your soul be ready for the mansions bright
And be washed in the blood of the Lamb?
Chorus

Lay aside the garments that are stained with sin
And be washed in the blood of the Lamb;
There's a fountain flowing for the soul unclean,
O be washed in the blood of the Lamb!
Chorus

Elisha A. Hoffman

OPHIUCHUS "THE SERPENT HOLDER"
CHAPTER 3 SECTION 2

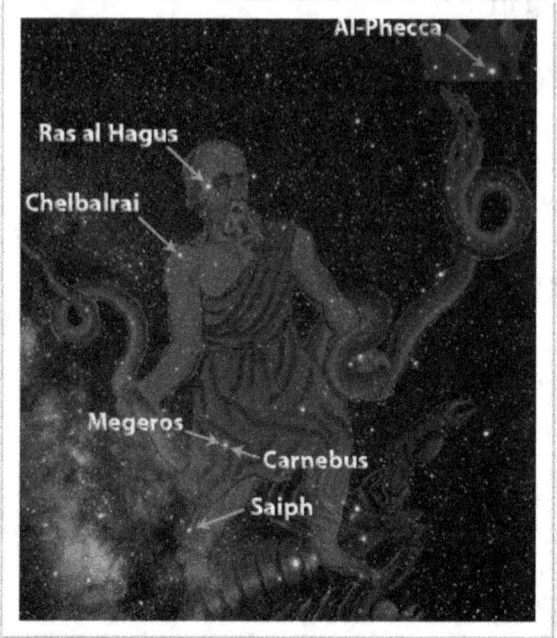

Ophiuchus is the first decan of Scorpio. From the standpoint of his positioning we find the stinger of Scorpio near one foot and the body of Scorpio being crushed by the other. This image pictures perfectly the proto-evangelism of Genesis 3:15, "I will put enmity between thee and the woman, and between thy seed and her seed; it shall bruise thy head, and thou shalt bruise his heel." In addition, the stars in his body and hands which hold Serpens, *the Serpent*, revealing the Ophiuchus as the great Restrainer of Satan in the Church Age; namely, the Holy Spirit. Let's look at these stars now.

> And now ye know what withholdeth that he might be revealed in his time. For the mystery of iniquity doth already work: only he who now letteth *will let,* until he be taken out of the way. And then shall that Wicked be revealed, whom the Lord shall consume with the spirit of his mouth, and shall destroy with the brightness of his coming: *Even him,* whose coming is after the working of Satan with all power and signs and lying wonders, And with all deceivableness of unrighteousness in them that perish; because

they received not the love of the truth, that they might be saved. And for this cause God shall send them strong delusion, that they should believe a lie: That they all might be damned who believed not the truth, but had pleasure in unrighteousness.[33]

Here we see that the Holy Spirit in the lives of the believers is the only power holding back the forces of darkness. When "He" is taken out of the way, the Satanic power and influence will be unleashed upon the earth in such a way that the world has yet to experience.

The Principal Stars

- Afeichus Hebrew and Arabic meaning, "**the serpent held.**"
- Chelbalrai is a star in Serpens, which in Arabic means "enfolding." He can't beat the Restrainer, so he enfolds himself around to restrict the Restrainer as much as he can.
- Saiph is a star in the right foot of Ophiuchus and means "bruised." What is significant about this star is that it is being stung by Scorpio, Ophiucus' other foot being on the head of Scorpio.
- Carnebus means "the wounding."
- Megeros, the same as Carnebus, means "contending."
- Ras al Hagus is a star in the head of Ophiuchus, which is Arabic and means "the head of him who holds." The name of this star in the Greek is Ophiuchus, and means "holder of the serpent."
- Al-Phecca is a star of the first magnitude in the Northern Crown, who's name in Arabic means "the shining." The Arabic name for the Northern Cross is *Al-Iclil*, and means an *ornament* or *jewel*.

From the Greek mythology some Christian authors have tried to attribute Ophiucus to Christ's attribute of the Great Physician. The mythology is as follows:

> But the Serpent-holder generally was identified with Ασκληπιος, Asclepios, or Aesculapius, whom King James I described as "a medicine after made a god" with whose worship serpents were always associated as symbols of prudence, renovation, wisdom, and the power of discovering healing herbs. Educated by his father Apollo,

---

[33] 2nd Thess. 2:6-12

or by the Centaur Chiron, Aesculapius was the earliest of his profession and the ship's surgeon of Argo. When the famous voyage was over he became so skilled in practice that he even restored the dead to life, among these being Hippolytus, of whom King wrote: "After his members were drawn in sunder by four horses, Esculapius at Neptune's request glued them together and revived him."[34]

During the time that John was on the Isle of Patmos, he wrote to the church at Pergamum mentioning that they dwelt where Satan's seat was (Rev. 2:13). The Temple in Pergamus was dedicated to the Greeks' four main deities: Zeus (the head of all the gods), Dionysius (the god of wine and drama), Athena (the god of wisdom in art and war), and Aesculapius (the god of healing). The Pergamum Church was accused by Christ as having held the doctrine of Balaam which can be attributed to compromise.

Where the idea derived of Ophiuchus being a great healer within the astrological record is impossible to identify. From the earliest Egyptian zodiacs (Esnah/Denderah) he is represented by an enthroned man wearing the head of a hawk—revealing that he is the enemy of the serpent. Although Ophiuchus could easily fit the attribute of Christ as Great Physician there is no evidence of it in star name etymology.

Satan wants to have a kingdom on this earth. His lust and greed for power is ever present. As he seeks to exalt his throne above the stars of God (Is. 14:13), so, he aims for "the Northern Crown" (Corona Borealis), the third decan of Libra. Although Christ, through a *just plan,* earned the crown, Satan, through an *unjust plan,* seeks to steal it.

Once the biblical story of God is taken into consideration, the crown can be identified as "the peculiar treasure of God," namely the Old (Ex. 19:5) and New Testament saints (Tit. 2:14); a truth which is signified by the four and twenty elders in John's Apocalypse (Rev. 4:4, 10; 5:8, 14; 11:16; 19:4) representing the 12 tribes of Israel and the 12 apostles of Christ (Rev. 21:12-14). No matter how hard Satan tries, he cannot touch God's property without permission, and even then, it is only for God's higher

---

[34] *Star Names, Their Lore and Meaning*, by Richard Hinckley Allen, Dover Publications Inc., Pg. 298, New York, original 1863, revised 1963.

good. Thus, in the heavenly portrayal of Ophiuchus he restrains Serpens from getting Corona Borealis.

The Bible says that "the mystery of iniquity doth already work: only he who now letteth *will let*, until he be taken out of the way. And then shall that Wicked be revealed, whom the Lord shall consume with the spirit of his mouth, and shall destroy with the brightness of his coming: *Even him*, whose coming is after the working of Satan with all power and signs and lying wonders, And with all deceivableness of unrighteousness in them that perish; because they received not the love of the truth, that they might be saved" (2nd Thess. 2:7-10). Here we see that the great Restrainer of the kingdom of Satan is none other than the third Person of the Godhead, namely, the Holy Spirit. No matter how much Satan has tried to destroy the kingdom of God, it has always been to no avail. Historically, as Satan persecuted the church through various means it only multiplied and grew.

In this world, in which Satan is the prince of the power of the air, if one seeks Christ in his heart, he is then drawn by the Holy Spirit to righteousness. The Holy Spirit is the great Sealer of the Church of God (Eph. 1:13; 4:30), meaning that he places the believer – at his heavenly birth – into the Church of the redeemed. This is the Church referred to by Christ when he said, "upon this rock (the truth of Christ's Messiah-ship revealed by the Father through the Agent of the Holy Spirit) I will build my church; and the gates of hell shall not prevail against it" (Matt. 16:15-18). Not only is the Holy Spirit the Restrainer for the collective body of Christ, namely the Church, but also for each individual believer. This is seen in Titus 2:14 "Who gave himself for us, that he might redeem us from all iniquity, and purify unto himself a peculiar people, zealous of good works."

The word *peculiar* is the Greek word περιουσιον, pronounced in English as per-ee-oo-see-on. The affixed preposition *peri* means *around* and the other portion of the word means *special*. So we see that Christians are special in the sense that they have a personal Bodyguard in the Holy Spirit of God. This trademark of peculiarity allows them to be zealous of good works in light of evil and troublesome times. The Holy Spirit walks guard around his chosen. What a comfort it is to know that nothing can come to a believer which isn't first allowed by God. Since we know that God loves us, we know that whatever is allowed into our realm has to first come through the permissive will of God. Therefore, whatever we face, we can handle it, because of the assurance that it is a part of God's plan.[35]

The stars in Ophiuchus are still speaking to us today. In this image of the invisible God, we see the wonderful attribute of the Holy Spirit as the Restrainer of evil. Christ reminded his disciples of the Holy Spirit's coming soon after he would bodily leave the earth, "I will not leave you comfortless: I will come to you" (Jn. 14:18). In this verse, it is significant to notice that the Greek word for *comfortless* is ορφανος, pronounced in English as *or-fan-os*, which is where we get our word *orphan*. Christ was saying, I will not leave you an orphan, but will send the Comforter (παράκλητη, pronounced *para-clay-tay* one called alongside), who is the Holy Spirit (Jn. 14:16, 26). How much more peculiar could we be to Christ, who first loved us? If that doesn't increase our faith, what will?

---

[35]Disclaimer: Christians can bring into their lives unnecessary evils, sins and circumstances that are contrary to the will of God. So, it is understood that the marks of peculiarity mentioned here refer to the righteous saints in God's will, not the unwilling and disobedient children of God.

The Comforter has Come

O spread the tidings 'round; wherever man is found,
Wherever human hearts and human woes abound;
Let ev'ry Christian tongue proclaim the joyful sound:
The Comforter has come!

Refrain: The Comforter has come! The Comforter has come!
The Holy Ghost from Heav'n, the Father's promise giv'n!
O spread the tidings 'round, wherever man is found:
The Comforter has come!

The long, long night is past; the morning breaks at last;
And hushed the dreadful wail and fury of the blast,
As o'er the golden hills the day advances fast!
The Comforter has come! Refrain

Lo, the great King of kings, with healing in His wings,
To ev'ry captive soul a full deliverance brings;
And through the vacant cells the song of triumph rings;
The Comforter has come! Refrain

O boundless love divine! How shall this tongue of mine
To wond'ring mortals tell the matchless grace divine—
That I, a child of hell, should in His image shine!
The Comforter has come! Refrain

By: Frank Bottome

HERCULES (THE MIGHTY)
CHAPTER 3 SECTION 3

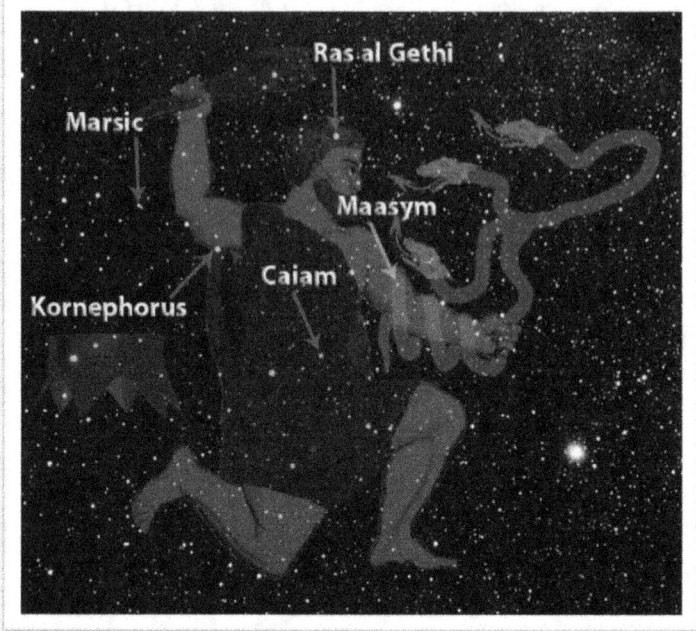

"For unto us a child is born, unto us a son is given: and the government shall be upon his shoulder: and his name shall be called Wonderful, Counsellor, <u>The mighty God</u>, The everlasting Father, The Prince of Peace" (Isaiah 9:6).

Hercules is the third decan of Scorpio and pictures Christ's triumph over Satan at Calvary. The statement "the mighty God" in Isaiah 9:6 is the Hebrew compound **el-Gibbor**, which may refer to Hercules, in that **Gibbor** is the Hebrew name for the constellation of Hercules.[36] Since the affixed article *el* is definite, it is possible that the inference behind the writing was that the prophesied individual of Isaiah 9:6 would be the incarnate of the heavenly **el-Gibbor** (Hercules), not in the sense that Greek mythology has attributed, but that which God created (Is. 40:26), numbered (Ps. 147:4), signified (Gen. 1:14), set (Ps. 19:4), and named (Ps. 147:4) within the confines of his astronomical blueprints.

---

[36] E.W. Bullinger, The Witness of the Stars, Kregel Classics, 1893 reprint.

Once all the facts are taken into view, there is little doubt that Hercules is another picture of the Genesis 3:15 *proto-evangelism*, or first Gospel announcement. Some have tried to build a case against Hercules, insinuating that it is not a part of the astronomical classics, by saying that it was invented by the Greeks. In spite of these modern arguments, Hercules dates back to antiquity in Egyptian hieroglyphics and star charts, as well as Phoenician and Indian Astrology antedating the early Greek era.

PRINCIPAL STARS

"Lift up your eyes on high, and behold who hath created these *things*, that bringeth out their host by number: he calleth them all by names by the greatness of his might, for that *he is* strong in power; not one faileth" (Isaiah 40:26). When we begin to believe what God has said about the stars in his Word, it is then that the intelligence behind the stars can speak to ours. It doesn't take God a lot of effort to call stars by their names, thus the illuminated truth within the statement "by the greatness of his might" (Isa. 40:26) is that the stars' names themselves reveal God's might, strength and power. This revelation is not done from the creative aspect only, but by means of an intelligence aspect. Hercules' principal stars expose this truth immensely.

- *El Giscale*, meaning "the strong."
- *Caiam*, meaning "treading under foot." - Luke 10:18-19 "And he said unto them, I beheld Satan as lightning fall from heaven. Behold, I give unto you power to tread on serpents and scorpions, and over all the power of the enemy: and nothing shall by any means hurt you." (See also Genesis 3:15)
- *Maasym*, meaning "the sin offering."
- *Kornephorus*, meaning "the branch." The Old Testament writers, in more places than one (Zech. 3:8; 6:12; Jer. 23:5-6; 33:15), prophesied that there was to come a Branch (Messiah) from the seed of Jesse (Is. 11:1-5). This prophecy was fulfilled in Christ and validated in the New Testament (Matt. 1:6-16; Rev. 5:5; 22:16). In Revelation 22:16 the Bible says, "I Jesus have sent mine angel to testify unto you these things in the churches. I am the root and the offspring of David, and the bright and morning star." The Greek word for "root" is ρἰζα and is the same underlying word in the Septuagint (LXX) for *stem* and *roots* in the Isaiah 11:1 prophecy. This signifies that Christ is the fulfillment of

43

the many Old Testament Branch prophecies (Ps. 80:15-17; Is. 11:10; 53:2; Jer. 23:5; 33:15; Zech. 3:8; 6:12; etc.).

- *Ras al Gethi* means the "head of him who bruises." This star pictures well the proto-evangelism of Genesis 3:15. "And I will put enmity between thee and the woman, and between thy seed and her seed; it shall bruise thy head, and <u>thou shalt bruise his heel</u>." Here we find that the coming seed would bruise the serpent's head. *Ras al Gethi* is located in the head of Hercules, and powerfully puts forth that he is the coming seed of Virgo that would take out the dragon. Fittingly, under Hercules' heel is Draco's head.

Remember, according to our model the image of Hercules has to be a derivative of the intelligence behind the stars' names. Thus we have a kneeling man with a club in one hand and a three headed dragon, named **Cerberus** in the other. In legend, **Cerberus** kept the gates of Hell.

The Descension of Christ
It is known that after Christ died he descended into the lower parts of the earth, taking the keys of death and Hell, symbolically defeating *Cerberus*, "the keeper of the gates of hell." Notice what the Bible says in Ephesians 4:8-10, "Wherefore he saith, When he ascended up on high, he led captivity captive, and gave gifts unto men. Now that he ascended, what is it but that he also descended first into the lower parts of the earth? He that descended is the same also that ascended up far above all heavens, that he might fill all things." Could it be that during this time of descending that Christ defeated Cerberus? Christ said in Revelation 1:18, "I am he that liveth, and was dead; and, behold, I am alive forevermore, Amen; and have the keys of hell and of death."

The Great Kneeling of Christ
The idea behind the kneeling demonstrates perfectly the humility of Christ. Philippians informs us that the mind of Christ was his taking on "the form of a servant, and was made in the likeness of men: And being found in fashion as a man, he humbled himself, and became obedient unto death, even the death of the cross" (Phil. 2:7-8). Thus, we see the great kneeling of our Hercules. In this wonderful image of the invisible God, we observe the secret to power is not in force, but in humility. The secret to rising is kneeling. In 2nd Corinthians 12:9, Christ told Paul, "My grace is sufficient for thee: for my strength is made perfect in weakness." Paul

said, "Most gladly therefore will I rather glory in my infirmities, that the power of Christ may rest upon me." James alarms us that, "God resisteth the proud, but giveth grace unto the humble" (4:6). Again in James 4:10 we read, "Humble yourselves in the sight of the Lord, and he shall lift you up." 1st Peter 5:6 tells us, "Humble yourselves therefore under the mighty hand of God, that he may exalt you in due time."

Humility was a necessary ingredient that Christ used to obtain victory over Satan's dominion of sin and death (Heb. 2:14). On Hercules' right side, we see Corona Borealis (Latin for the Northern Crown) which Serpens aims to take. This crown was temporarily set aside after the Garden of Eden Story until Christ's great conflict. Through this struggle, Corona Borealis was on trial to prove if indeed Christ was "all and in all" (Col. 3:11). In the end of Christ's competition, we see his triumph (resurrection) and exaltation (ascension to the right hand of God), proving his Lordship and pre-eminence. Notice, during the time that *Corona Borealis* was set aside, Christ took upon the form of a servant (or slave δουλος) in order to pay the sin debt of humanity and crush the dominion Satan had over humanity's eternal separation from God (spiritual death).

Not only did Calvary provide salvation, the source of cleansing and justification for the believer, but it was indeed the mighty battle axe of Hercules, which defeated Satan's dominion (κρατος) over death (θανατος) as mentioned in Hebrews 2:14. "Forasmuch then as the children are partakers of flesh and blood, he also himself likewise took part of the same; that through death he might destroy him that had the power of death, that is, the devil." This dominion goes all the way back to the Garden of Eden, when God commanded Adam not to eat of the tree of the knowledge of good and evil (Gen. 2:17). When the *fall of man* took place, the human race became estranged from God and eternally separated.

God, having a revolution on his hands with Lucifer and the angles prior to the Garden of Eden, would now have a rebellion with man. Deciding to intervene, God made a plan to save the human race. This plan of intervention was one of love. In John 3:16, Christ said, "For God so loved the world, that he gave his only begotten Son, that whosoever believeth in him should not perish, but have everlasting life."

To perform his act of love, Christ continued down his "despised and rejected of men; a man of sorrows, and acquainted with grief" prophecy in Isaiah 53. Even at the Cross we see him going down further and further. He was handed vinegar when he was thirsty, mocked and railed on the thieves and some of the onlookers, "Likewise also the chief priests mocking *him,* with the scribes and elders, said, He saved others; himself he cannot save. If he be the King of Israel, let him now come down from the cross, and we will believe him. He trusted in God; let him deliver him now, if he will have him: for he said, I am the Son of God. The thieves also, which were crucified with him, cast the same in his teeth. (Matt. 27:41-44). He even went lower as he cried, , "My God, My God, why hast thou forsaken me" (Matt. 27:46). Here, Christ felt the eternal separation and utter forsaking of God in a human body for humanity, as he who knew no sin, was made sin for us (2nd Cor. 5:21).

Christ had to go down, to redeem fallen man. In Matthew 12:40, Christ said, "For as Jonah was three days and three nights in the whale's belly; so shall the Son of man be three days and three nights in the heart of the earth." After Christ died on Golgotha's Hill, it was then that the invisible God went to Hell for the sins of the world (κοσμος). This is called the Descension of Christ.

This heart-wrenching revelation, teaches that the greatness of Christ's life and suffering at Calvary was not just measured in height at the Resurrection and Ascension, but also in depth. Therefore, everyone who wants to know Christ must first know his *great kneeling* and his *great descension.* The King must be envisioned laying down his crown, being tabernacled in human flesh, becoming a slave, sacrificing his life, seen eternally forsaken of the Father, paying the sin debt of humanity, and descending into Hell. What more manifestation and proof that "God so loved the world" (John 3:16)? Christ as Lamb and Lion are wonderfully balanced in the message of Hercules.

46

Sagittarius "The Victorious One"
Chapter 4 Section 1

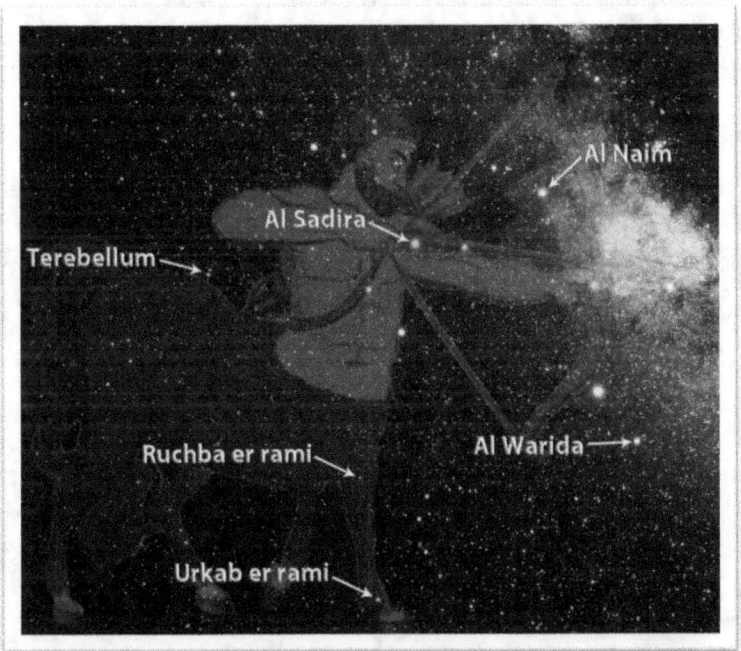

Sagittarius, like the other 11 main constellations, has three decans; *Lyra*, *Ara*, and *Draco*. In this heavenly portrait, we see a dual-natured Centaur, signifying Deity and the incarnate Christ. It was because of Christ's dual nature (man and God) that Satan would lose the war with God. Scorpio having affixed his stinger to the foot of Ophiuchus (signifying Calvary), did not realize that the death, burial, descension, resurrection and ascension of Christ would through time, come from nowhere and smite the heart of the arch enemy with a dart. This is easily seen from Sagittarius' arrow aimed at Scorpio. What Satan thought was victory became his demise, forcing him to lose at Calvary, the human race that he once subjugated in the Garden of Eden. The stars of Sagittarius reveal a panorama of the gospel story found only in *Biblical Astrology*. Here, the coming seed defeats the enemy and is, thus, the Victorious One over sin, the grave, death, and Hell. Through this victory, he both conquered the enemy and provided a blood bought redemption for the subjugated race.

The tale of Sagittarius is typed out in Psalm 45. "My heart is inditing a good matter: I speak of the things which I have made touching the king: my tongue *is* the pen of a ready writer. Thou art fairer than the children of men: grace is poured into thy lips: therefore God hath blessed thee for ever. Gird thy sword upon *thy* thigh, O *most* mighty, with thy glory and thy majesty. And in thy majesty ride prosperously because of truth and meekness *and* righteousness; and thy right hand shall teach thee terrible things. Thine arrows *are* sharp in the heart of the king's enemies; *whereby* the people fall under thee" (Psalms 45:1-5). In this Epithalamium,[37] we see the Bridegroom's victory and the glorious Bride of the Redeemed. "Kings' daughters *were* among thy honourable women: upon thy right hand did stand the queen in gold of Ophir" (Ps. 45:9).

PRINCIPAL STARS

- *Al-kaus* – "the arrow."
- *Al Naim* – the brightest star in Sagittarius, which in the Hebrew means "the gracious one," a complement of our Lord and Savior.
- *Al Sadira* – "who strives."
- *Al Shuala* – "the dart."
- *Al Warida* – "who comes forth."
- *Ruchba er rami* – "the riding of the bowman."
- *Urkab er rami* – "the bowman, the rider."
- *Croton* – "the purchaser." Here is a star that signifies the rider, as the Redeemer. Christ is our great Redeemer.
- *Terebellum* – "sent forth swiftly."

The stars in Sagittarius are still speaking to us today. They are telling us of the Gracious One, Jesus Christ. One who had the authority at the time of his death to "pray to my Father, and he shall presently give me more than twelve legions of angels" (Matt. 26:53), but chose instead to die in our place "that we might be made the righteousness of God in him" (2nd Cor. 5:21). Sagittarius shows humanity that their victory in this life must exist in the unmerited favor and love of God.

---

[37] Noah Webster said an "EPITHALA'MIUM is a nuptial song or poem in praise of the bride and bridegroom, and praying for their prosperity. The forty-fifth Psalm is an epithalamium to Christ and the church."

"Amazing Grace"

Amazing Grace how sweet the sound—
That saved a wretch like me!
I once was lost, but now I'm found,
Was blind but now I see.

Twas grace that taught my heart to fear,
And grace my fears relieved;
How precious did that grace appear
The hour I first believed!

Thru many dangers, toils and snares
I have already come;
Tis grace hath brought me safe thus far,
And grace will lead me home.

When we've been there ten thousand years,
Bright shining as the sun,
We've no less days to sing God's praise
Than when we've first begun.

John Newton (1725-1807)

LYRA "THE HARP"
CHAPTER 4 SECTION 2

Lyra is the first decan in the Constellation of Sagittarius. In the Zodiac of Denderah, this constellation's name is *Fent-kar*, which means "the serpent ruled." The word *ruled* signifies "confined, bound, and imprisoned." Lyra is figured as an eagle carrying a harp. Of course, the pictures in the heavens are not derived by literal images in the heavens, but by the names of the stars themselves (see the four principals in the introduction). The eagle signifies the upward praise of the harp. Below, we see the principal stars in Lyra and thus, the origin of the heavenly image.

PRINCIPAL STARS
- The main star in Lyra is *Vega*. Vega is one of the most sparkling diamonds in the night sky. Vega means "he shall be exalted." Psalms 21:12-13 tells us, "Therefore shalt thou make them turn their back, *when* thou shalt make ready *thine arrows* upon thy strings against the face of them. Be thou exalted, LORD, in thine own strength: *so* will we sing and praise thy power."
- *Shelyuk* (Arabic Al Nesr) means "an eagle."
- *Sulaphat* means "springing up, or ascending."

The harp is signified as praise. Notice the harps' role in the following verses:

- Revelation 5:8, "And when he had taken the book, the four beasts and four *and* twenty elders fell down before the Lamb, having every one of them harps, and golden vials full of odours, which are the prayers of saints."
- Revelation 14:12, "And I heard a voice from heaven, as the voice of many waters, and as the voice of a great thunder: and I heard the voice of harpers harping with their harps."
- Revelation 15:2, "And I saw as it were a sea of glass mingled with fire: and them that had gotten the victory over the beast, and over his image, and over his mark, *and* over the number of his name, stand on the sea of glass, having the harps of God."
- Revelation 19:1-7, "And after these things I heard a great voice of much people in heaven, saying, Alleluia; Salvation, and glory, and honour, and power, unto the Lord our God: For true and righteous *are* his judgments: for he hath judged the great whore, which did corrupt the earth with her fornication, and hath avenged the blood of his servants at her hand. And again they said, Alleluia. And her smoke rose up for ever and ever. And the four and twenty elders and the four beasts fell down and worshipped God that sat on the throne, saying, Amen; Alleluia. And a voice came out of the throne, saying, Praise our God, all ye his servants, and ye that fear him, both small and great. And I heard as it were the voice of a great multitude, and as the voice of many waters, and as the voice of mighty thunderings, saying, Alleluia: for the Lord God omnipotent reigneth. Let us be glad and rejoice, and give honour to him: for the marriage of the Lamb is come, and his wife hath made herself ready."

When we think of *Alleluia* in Revelation 19, it is important to know that the setting of the story is at the beginning of the battle of Armageddon. The word *alleluia* is a Greek transliteration of two Hebrew words, *Alel*, and *Yah*. **Alel** means praise, and **Yah** means Lord, thus the compound word means *praise the Lord*. This particular Alleluia is the announcement of Christ's vengeance upon the forces of evil. Christ will, at this time, be exalted in another *Day of the Lord*. This Day begins when Christ descends from Heaven with a sharp two edged sword, destroys the

51

enemies of darkness, and binds Satan in the bottomless pit for one thousand years. Thus, we see *Fent-kar,* meaning "the serpent ruled or contained," which is the Coptic name for Lyra. It is at this time that the harp takes its flight high into the heavens, and the entire universe pays homage to the Lamb of God. What greater inspiration could we receive from such an image, yet it is the very plot of the prophetic story found within Scripture.

With that blessed hope before us,
Let no harp remain unstrung;
Let the mighty advent chorus
Onward roll from tongue to tongue:
Christ is coming! Christ is coming!
Come, Lord Jesus, quickly come!

John Ross Macduff, 1853

Alleluia!!

Draco "The Dragon"
Chapter 4 Section 3

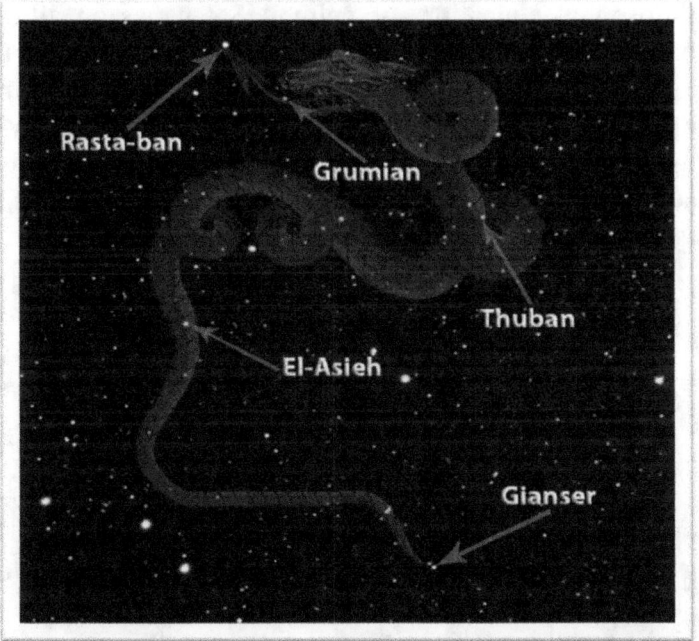

Draco is a decan of Sagittarius. He also is mentioned in Job, the oldest book of the Bible. In Job 26:13, Job declared that God's "hand formed the crooked Serpent." The Septuagint renders *serpent* as δράκοντα, pronounced in English as *draconta*, otherwise known as Draco. Looking up into the night sky, one will not see an image of a serpent. However, if he studied the names of the stars in this constellation, he could receive the heavenly portrait and the story that underlies it. The Bible says in Isaiah 40:26 "Lift up your eyes on high, and behold who hath created these *things*, that bringeth out their host by number: **he calleth them all by names** by the greatness of his might, for that *he is* strong in power; not one faileth." There are five major stars in Draco which depict paramount doctrine within the gospel. This doctrine is called *the Personal Devil*.

The doctrine of *a Personal Devil* is often overlooked or misunderstood. Although most Christians believe in a *Personal Devil*, it is seldom seen in their practice. Thus, the Devil has become a *principle of evil* instead of the intended *prince of evil*. In fact, it has become quite popular in many circles today to teach about the Devil with the "d" left off, denying his real existence, making only "evil" the arch enemy. This is seen when Christians fail to acknowledge fingerprints of Satan behind every sin, broken home, and debauched life. Whether it be in religion, science, psyche, philosophy, education, or culture; Satan has an all-out assault against humanity. It is important therefore to believe that all the evil in the world has a motivating source behind it. Satan has strategically positioned himself in all the high grounds of the world. Meanwhile, Christianity has been "chompin' at the bit" treating the problems, but leaving the root cause of the chaos untouched and unnoticed. Draco's presence is seen within the majority of the stars. One writer said,

> With vast convolutions Draco holds
> Th' ecliptic axis in his scaly folds.
> O'er half the skies his neck enormous rears,
> And with immense meanders parts the Bears.[38]

Principal Stars

The first star in Draco is *Thuban*, which means "the subtle." This star is describing the sly and artful way the Devil works. It was through his subtlety that he was able to deceive Adam and Eve. Notice what the Bible warns in 2nd Corinthians 11:3a "But I fear, lest by any means, as the serpent beguiled Eve through his subtilty." The Bible also declares in Genesis 3:1 "Now the serpent was more subtil than any beast of the field which the LORD God had made." Notice, it was the subtleness of the serpent that deceived Adam and Eve during *the Fall of Man* in the Garden of Eden. Adam and Eve may have brought the first sin into the human race, but it was Satan who authored and originated sin. People often ask why there is so much evil in the world, in an attempt to make a disclaimer against the existence of God. This enigma has troubled many people through the years. This question is a paradox in that it proposes that the person making this accusation acknowledges evil. If they acknowledge evil, then they must acknowledge righteousness. Deduction: To know that there is a righteous God, we must have knowledge of God's co-equal opposite, Satan. If God is Light, then Satan must be darkness. Since evil exists, Satan exists; and since Satan exists, God must exist. Here we see the extreme powers in the universe, and the souls of men in the balance.

---

[38] *The Botanic Garden*, by Erasmus Darwin (1731-1802), A Poem in Two Parts, Part I "the Economy of Vegetation," By Scholar Select, reprint, no date.

Since Satan imitates a replica of God in many respects, it is important for the believer to gather wisdom, prudence and discretion from God's Word.

The second star in this constellation is *Rasta-ban*, "head of the subtle." As Christ is the head of the Church, Satan is the head of his kingdom. He is the leader of all the tricksters and conniving liars that get over on everybody. His leaders are "the rulers of the darkness of this world" (Eph. 6:12) and although they are subtle, they are exposed by the Word of God. Their influence derives from their master and father, the Devil. Jesus acknowledged the Devil's position of authority when he told the Pharisees in John 8:44, "Ye are of your father the devil, and the lusts of your father ye will do. He was a murderer **from the beginning**, and abode not in the truth, because there is no truth in him." The Pharisees were so high up *the social ladder* that they were trusting in their religion to save them and not their religion's God. They believed in a works salvation and denied Christ's deity. They wanted signs, but not the Sign of Old Testament prophecy. They wanted wonders, but not the Wonder of the universe. How sad, but how true it is today, that many people just want God for what they can get out of him, but not for him.

The third star in this constellation is *Grumian*, which means "the deceiver." "And the devil that deceived them was cast into the lake of fire" (Rev. 20:10a). There is not a more proper word that better depicts the Devil than *deceiver*. Everyone has been deceived at one time or another. No one has ever enjoyed being deceived, yet at this moment everyone in the world is deceived by something or someone; it just hasn't been revealed yet. Likewise, Satan never tells us the end results. He wants to sell us his bag of goods, never letting us know all the terms and conditions. From addictions to crime, from bad relationships to disease, his terms and conditions are never spelled out in black and white. Thus, he is the grand-master of deception.

The fourth star in this constellation is *El-Asieh*, which means "the humbled." There will come a time when the Devil will be bound for a thousand years during the millennial reign of Christ. "And he laid hold on the dragon, that old serpent, which is the Devil, and Satan, and bound him a thousand years" (Rev. 20:2). What a humbling that will be for him, and what a blessing that will be for us. Can you imagine the world without the Devil? Speaking of this time, Isaiah says "The wolf also shall dwell with

55

the lamb, and the leopard shall lie down with the kid; and the calf and the young lion and the fatling together; and a little child shall lead them" (Is. 11:6). Concerning life "There shall be no more thence an infant of days, nor an old man that hath not filled his days: <u>for the child shall die an hundred years old</u>; but the sinner *being* an hundred years old shall be accursed" (Is. 65:20). What a blessed time the earth will have during the Millennial Reign of Christ. However, we do not have to wait till the Millennial Reign to see Satan humbled. Even now, every time a Christian gets the victory over Satan, the Devil is humbled. Every time someone gets victory over their addiction, the Devil is humbled. Every time someone is converted and turned into a decent person that loves and serves God, the Devil is humbled. The thousand years of humbling will not lead him to repentance, but fuel his fire to go out into the world and deceive the nations once more. This is called his *Final Rebellion* and introduces perfectly our last star called *Gianser*.

The fifth star in this constellation is *Gianser*, "the punished enemy." After all of Satan's humbling experiences, his pride gets him once again and leads him to fight against the Lord one last time in his *Final Rebellion*. Satan's final humbling will be humiliation as he will be cast into the lake of fire for ever and ever. "And the devil that deceived them was cast into the lake of fire and brimstone, where the beast and the false prophet are, and shall be tormented day and night for ever and ever" (Rev. 20:10). When this event takes place, the time-clock as we know it will stop forever. From that day on, we will live in perfect harmony with God. It will be a state of eternal love and righteousness, without the presence of evil anymore. There will be no more suffering, no more disease, no more hospitals, no more addictions, no more evil and no more Devil! What an eternal experience that will be! The only way to know that we can experience Heaven as our eternal destiny is if we have turned to the finished work of Christ to have Satan's power broken in our lives. Notice what Jesus said to Paul in Acts 26:18, "To open their eyes, and to turn them from darkness to light, and from the power of Satan unto God, that they may receive forgiveness of sins, and inheritance among them which are sanctified by faith that is in me."

Recap: The stars in Draco are still speaking to us today. They are a constant reminder of the Great War between good and evil, and Satan and God. These stars validate the teachings of the Word of God concerning

56

our arch enemy, and remind us to never underestimate "that old serpent" (Rev. 20:2). Throughout *Biblical Astrology*, you will see the repetitious pattern of the stars concerning God's arch enemy.

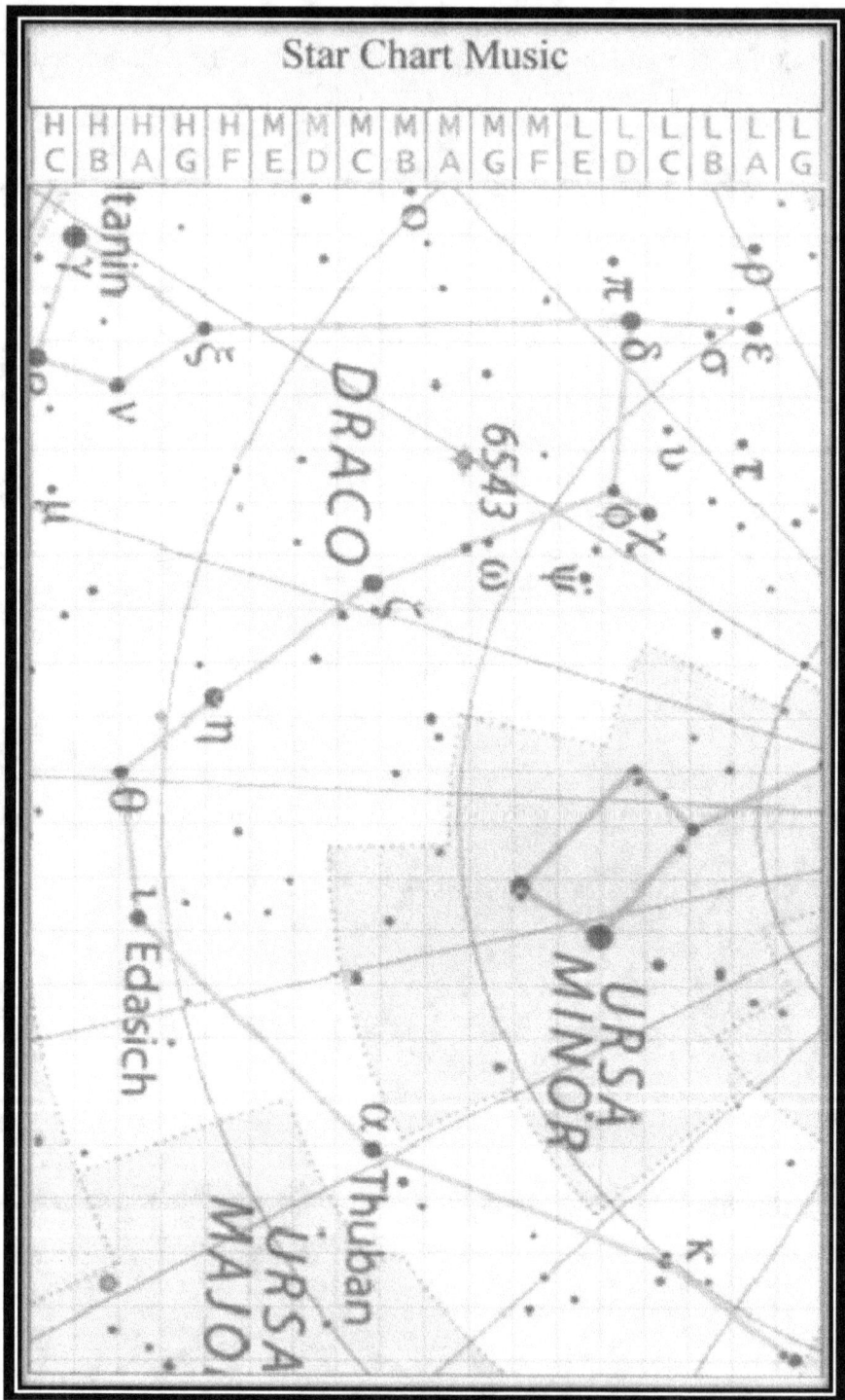

ARA "THE BURNING ALTAR"
CHAPTER 4 SECTION 4

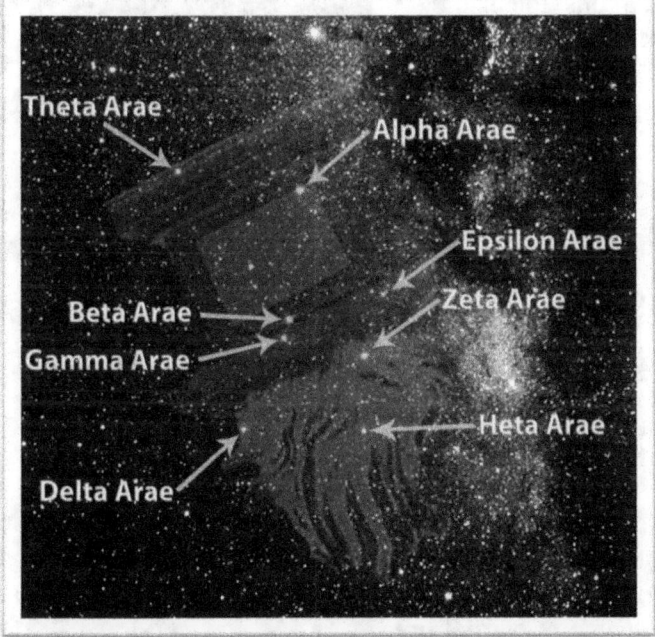

Because of the precession of the equinoxes, Ara is no longer visible from the northern latitudes, but deep in the southern hemisphere's night sky shines this brilliant and glimmering constellation called Ara. It is the third decan of Sagittarius and, because of the precession of the equinoxes, no longer visible from the northern latitudes. Having previously learned about Draco, Ara confirms the finality of Satan. The Arabic name for this constellation is pronounced in English as *Al Mugamra*, which means "completing, finishing, or finality." Ara connects directly in the transliterated Hebrew words *mara*, meaning "bitter or curse," and *aram* meaning "utter destruction." This constellation faces downward and is one of the southernmost constellations. Ara's fires burn downward into the Abyss of "outer darkness." Christ said in Matthew 8:12, "But the children of the kingdom shall be cast out into outer darkness: there shall be weeping and gnashing of teeth."

In the Zodiac of Dendera, Ara is under a different representation even though the story remains the same. At Dendera, we have an enthroned deity (picturing Christ) wielding a flail, an instrument for thrashing or beating corn from the ear. This figure is set over the infamous Jackal (picturing Satan). Combined, these images appear to teach the finality of Satan in the lake that burns with fire and brimstone. Notice the following New Testament verses:

- Jude 1:7, "Even as Sodom and Gomorrah, and the cities about them in like manner, giving themselves over to fornication, and going after strange flesh, are set forth for an example, suffering the vengeance of eternal fire."
- 2nd Thessalonians 1:8, "In flaming fire taking vengeance on them that know not God, and that obey not the gospel of our Lord Jesus Christ."
- 1st John 3:8-10, "He that committeth sin is of the devil; for the devil sinneth from the beginning. For this purpose the Son of God was manifested, that he might destroy the works of the devil. Whosoever is born of God doth not commit sin; for his seed remaineth in him: and he cannot sin, because he is born of God. In this the children of God are manifest, and the children of the devil: whosoever doeth not righteousness is not of God, neither he that loveth not his brother."

Here in this magnificent portrait of Satan's doom, we are reminded of the victory we have in Jesus Christ, who has saved us from the "everlasting fire, prepared for the devil and his angels" (Matt. 25:41). Those under the redemptive powers of Christ will not partake of this punishment with Satan's kingdom. "But the fearful, and unbelieving, and the abominable, and murderers, and whoremongers, and sorcerers, and idolaters, and all liars, shall have their part in the lake which burneth with fire and brimstone: which is the second death" (Revelation 21:8). In Revelation 20:10, we see the final doom of Satan as he is "cast into the lake of fire and brimstone, where the beast and the false prophet *are*, and shall be tormented day and night forever and ever."

"Victory in Jesus"

I heard an old, old story,
How a Savior came from glory,
How He gave His life on Calvary
To save a wretch like me:
I heard about His groaning,
Of His precious blood's atoning,
Then I repented of my sins
And won the victory.

Chorus
O victory in Jesus,
My Savior, forever!
He sought me and bought me
With His redeeming blood;
He loved me ere I knew Him,
And all my love is due Him—
He plunged me to victory
Beneath the cleansing flood.

(E.M. Bartlett 1939)

Capricornus "The Goat"
Chapter 5 Section 1

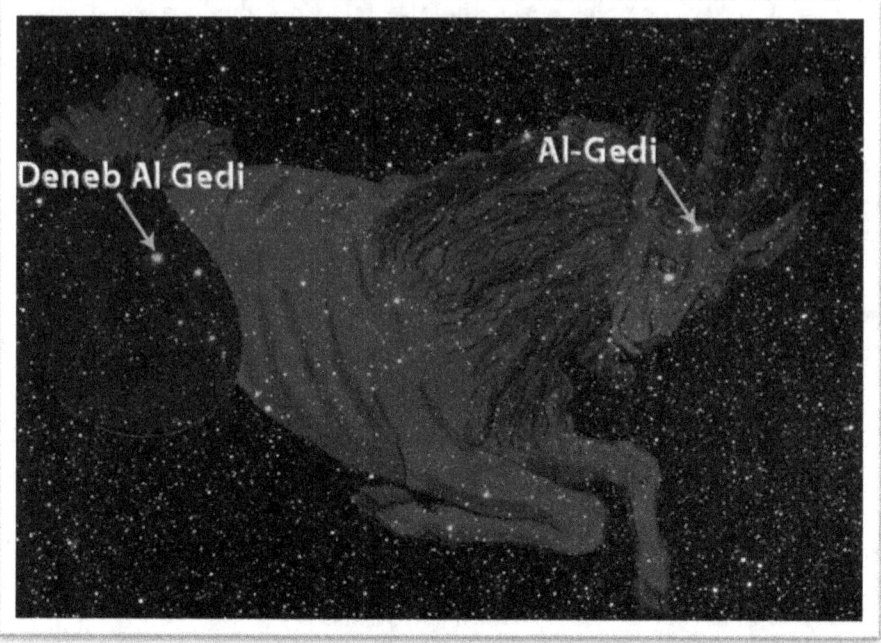

The Hebrew name for the constellation of Capricornus is Gedi, meaning *the sacrifice cometh*, or *the coming sacrifice*. The meaning is referring to the cutting off of the Messiah mentioned in Daniel 9:26. In the Egyptian Zodiacs of *Denderah* and *Esneh*, Capricornus is pictured as a half goat and half fish. Its Egyptian name is pronounced in English as *Hu-penius*, which means *the place of the sacrifice*. Capricornus has three decans; *Sagitta*, *Aquila*, and *Delphinus*. This constellation and its decans add an important chapter to Biblical Astrology.

Principal Stars
- *Al-Gedi* means "the kid," and is where the image of a goat comes from.
- *Deneb Al Gedi* means "the sacrifice cometh." Here, we see that the world was to be expecting an incarnation of Deity.
- *Ma Asad* means "the slaying." This star tells what the slaying produces; which is the lively fish (symbolizing the resurrection and church of the redeemed).

THE TWO-FOLD IMAGE

The image of this constellation is two-fold. One part is of a dying goat and the other is of a lively fish. This image is pictured in the Old Testament sin offering and atonement of the scapegoat in Leviticus 16. In the book of Leviticus, we learn of the priest's duty concerning the scapegoat.

> And Aaron shall cast lots upon the two goats; one lot for the LORD, and the other lot for the scapegoat. And Aaron shall bring the goat upon which the LORD'S lot fell, and offer him *for* a sin offering. But the goat, on which the lot fell to be the scapegoat, shall be presented alive before the LORD, to make an atonement with him, *and* to let him go for a scapegoat into the wilderness.... And he shall take of the blood of the bullock, and sprinkle *it* with his finger upon the mercy seat eastward; and before the mercy seat shall he sprinkle of the blood with his finger seven times. Then shall he kill the goat of the sin offering, that *is* for the people, and bring his blood within the vail, and do with that blood as he did with the blood of the bullock, and sprinkle it upon the mercy seat, and before the mercy seat: And he shall make an atonement for the holy *place*, because of the uncleanness of the children of Israel, and because of their transgressions in all their sins: and so shall he do for the tabernacle of the congregation, that remaineth among them in the midst of their uncleanness. ...And when he hath made an end of reconciling the holy *place*, and the tabernacle of the congregation, and the altar, he shall bring the live goat: And Aaron shall lay both his hands upon the head of the live goat, and confess over him all the iniquities of the children of Israel, and all their transgressions in all their sins, putting them upon the head of the goat, and shall send *him* away by the hand of a fit man into the wilderness: And the goat shall bear upon him all their iniquities unto a land not inhabited: and he shall let go the goat in the wilderness...And the bullock *for* the sin offering, and the goat *for* the sin offering, whose blood was brought in to make atonement in the holy *place*, shall *one* carry forth without the camp; and they shall burn in the fire their skins, and their flesh, and their dung. (Lev. 16:8-27)

The sin offering is a picture of Christ's sacrifice, showing a sufficient and worthy price. The scapegoat fleeing and freed into the wilderness symbolizes the eternal atonement of sin. At the time of Christ's crucifixion, it was a custom for the Jews to release a prisoner at the Passover (Matt. 27:15; Mk. 15:6; Lk. 23:17; Jn. 18:39); this stemmed from the scapegoat ceremony found in Leviticus 16. Thus, Barabbas became the world's symbol for the sinner's atonement,[39] and the sacrifice of Christ, the world's propitiation.[40]

CHRIST'S ETERNAL ATONEMENT

In the verses below, we can see eternal atonement for sin, which only Christ can offer to sinners.

- Psalms 103:12, "As far as the east is from the west, *so* far hath he removed our transgressions from us."
- Isaiah 43:25, "I will not remember thy sins."
- Jeremiah 31:34, "I will remember their sin no more."
- Jeremiah 50:20, "I will pardon them whom I reserve."
- Micah 7:18, "God pardoneth iniquity, and passeth by the transgression of the remnant of his heritage?"
- 1 John 1:7, "the blood of Jesus Christ his Son cleanseth us from all sin."

So, we see that God atones for our sin through Christ. It is because of this that sinners understand *the Doctrine of Grace*,[41] because it should have been us on the cross paying for our sins and then facing the judgment of God, "but God so loved the world." It is because of his sacrifice that we (the transgressors) are able to go free.

---

[39] Atonement is the reconciliation of the sinner by God's satisfaction for his sin in the personal suffering of Jesus Christ.

[40] Propitiation is the appeasing of God's wrath toward sin. (See Is. 53:6)

[41] Grace is God's favor upon unworthy sinners. It is understanding God's free unmerited love toward us. Through this comprehension, a divine influence rests upon the believer, thus, motivating man's will to spring forth with good works. This influence (grace) also becomes a motivating force and tool in the life of every believer to transform his mind, reform his life into the image of Christ, and perform good works from his new outlook on life.

Furthermore, in this heavenly portrait of Capricornus, we not only see the truth of the two goats mentioned in Leviticus, and the Jewish custom of letting the guilty go free, but we also see a beautiful New Testament truth. The sacrificed goat is a picture of Christ being made sin for us (II Cor. 5:21), as the LORD hath "laid on him the iniquity of us all" (Is. 53:6); however, the fishtail symbolizes his victorious resurrection. Christ would be killed and then go free, because he needed no atonement, thus, qualifying him for a victorious resurrection. The story of Christ's resurrection is the backbone of Christianity. The crucifixion and resurrection of Christ are the qualifying ingredients for sinners to have atonement and justification in the eyes of God. So yes, Christ is our sacrificial Goat, but he is also our resurrected Fish! He is the firstfruits that should come back from the dead, thus, believers are raised to walk in newness of life.[42]

THE ICHTHUS

In the time of the early church, there was an affixed sign of a fish given by the believers and for believers. Fish is the Greek word, pronounced in English as *ic-thus*. Some early church historians have said that the Ichthus was a symbol used under persecution times; whereby, drawing or showing it, allowed the other party to know it was okay to be in fellowship with this person or group. Usually, two people would be facing each other and one would draw an arch on the ground and the other individual on the other side would draw one, completing the sign of a fish. This sign of Ichthus signified one who is alive in Christ, having partaken of his atoning sacrifice and resurrection. The Scriptures illustrate this truth in Romans very well.

---

[42] Therefore we are buried with him by baptism into death: that like as Christ was raised up from the dead by the glory of the Father, even so we also should walk in newness of life. (Rom. 6:4)

Therefore we are buried with him by baptism into death: that like as Christ was raised up from the dead by the glory of the Father, even so we also should walk in newness of life. For if we have been planted together in the likeness of his death, we shall be also *in the likeness of his* resurrection: Knowing this, that our old man is crucified with *him*, that the body of sin might be destroyed, that henceforth we should not serve sin. For he that is dead is freed from sin. Now if we be dead with Christ, we believe that we shall also live with him: Knowing that Christ being raised from the dead dieth no more; death hath no more dominion over him. For in that he died, he died unto sin once: but in that he liveth, he liveth unto God. Likewise reckon ye also yourselves to be dead indeed unto sin, but alive unto God through Jesus Christ our Lord. (Romans 6:4-11)

Again in Galatians, we see the same truth. "I am crucified with Christ: nevertheless I live; yet not I, but Christ liveth in me: and the life which I now live in the flesh I live by the faith of the Son of God, who loved me, and gave himself for me" (Galatians 2:20). It is because he died for our sins and resurrected that we live and have life. Dear reader, are you trusting in the death and resurrection of Jesus Christ?

"He Lives"

I serve a risen Savior, He's in the world today;
I know that He is living, whatever men may say;
I see His hand of mercy, I hear His voice of cheer,
And just the time I need Him He's always near.

Chorus: He lives, He lives, Christ Jesus lives today!
He walks we me and talks with me along life's narrow way.
He lives, He lives, salvation to impart!
You ask me how I know He lives? He lives within my heart.

In all the world around me I see His loving care,
And tho' my heart grows weary I never will despair;
I know that He is leading thro' all the stormy blast,
The day of His appearing will come at last. Chorus

Rejoice, rejoice, O Christian, lift up your voice and sing,
Eternal hallelujahs to Jesus Christ the King!
The Hope of all who seek Him, the Help of all who find,
None other is so loving, so good and kind. Chorus

Alfred Henry Ackley (1933)

AQUILA AND SAGITTA
CHAPTER 5 SECTION 2

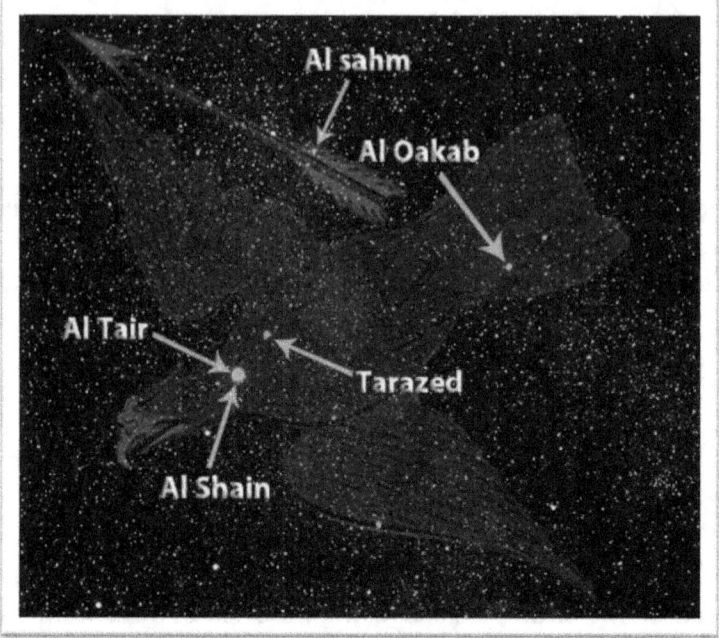

The Eagle and the Arrow are another testament of the Genesis 3:15 *proto-evangelism*. "And I will put enmity between thee and the woman, and between thy seed and her seed; it shall bruise thy head, and thou shalt bruise his heel" (Gen. 3:15). In Aquila, the eagle is falling, having been shot by Sagitta. Frances Rolleston said that Sagitta was "anciently said to be the Arrow that slew the Eagle."

The eagle is a symbol of Deity. The four Gospels beautifully depict the character and nature of Jesus Christ. Matthew depicts him as **King**, Mark as **Servant**, Luke as **Man**, and John as **God**. In the Old Testament, Christ is introduced by the word "behold" four times: a. "behold the King" (Zech. 9:9); b. "behold my servant" (Is. 42:1); c. "behold the man" (Zech. 6:12); and d. "behold your God" (Is. 40:9). The *early church* caught the spirit of the gospels and affixed *artwork* to each of them. Matthew gave Christ the symbol of a *lion* **(king)**; Mark, the symbol of an *ox* **(servant)**; Luke, the symbol of *man* **(son of man)**; and John, the symbol of an *eagle* **(God)**. These symbols are also used with the cherubim in Ezekiel 1:10

and Revelation 4:7. Thus, in Sagitta and Aquila, we see the Eagle, a symbol of deity, killed by piercing.

PRINCIPAL STARS[43]

Remember, our constellations are not derived through mysticism or connect-the-dot method, but by biblical principles given from Scripture. All stars have a name given to them by God. "He telleth the number of the stars; he calleth them all by *their* names" (Psalm 147:4). "Lift up your eyes on high, and behold who hath created these *things*, that bringeth out their host by number: he calleth them all by names by the greatness of his might, for that *he is* strong in power; not one faileth" (Isaiah 40:26). Below, we will visit these stars in *Aquila* and *Sagitta*.

- *Al Tair* means "*the wounded.*" Isaiah 53:5, "But he *was* wounded for our transgressions, *he was* bruised for our iniquities: the chastisement of our peace *was* upon him; and with his stripes we are healed."
- *Al Cair* means "*the piercing.*" Psalms 22:16, "For dogs have compassed me: the assembly of the wicked have enclosed me: they pierced my hands and my feet." John 19:37, "And again another Scripture saith, They shall look on him whom they pierced."
- *Al Oakab* means "*wounded in the heel.*" Genesis 3:15, "And I will put enmity between thee and the woman, and between thy seed and her seed; it shall bruise thy head, and thou shalt bruise his heel."
- *Al Shain* means "*the bright.*" Revelation 22:16, "I Jesus have sent mine angel to testify unto you these things in the churches. I am the root and the offspring of David, *and* the bright and morning star."

Knowing the story of Christ in relation to the astronomical names of the stars produces a mental spark between the Bible and the astrological record of the stars.

"Lift up your eyes on high, and behold who hath created these *things*, that bringeth out their host by number: he calleth them all by names by the greatness of his might, for that *he is* strong in power; not one faileth." (Is. 40:26)

---

[43] Ibid.

DELPHINUS "THE DOLPHIN"
CHAPTER 5 SECTION 3

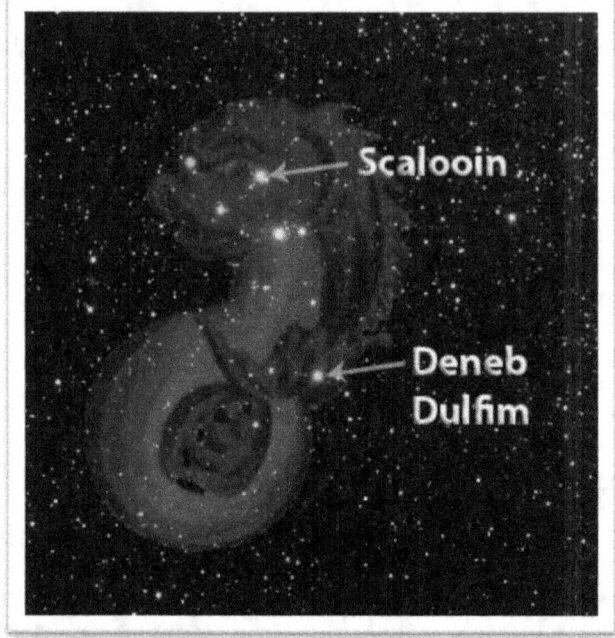

In this small constellation, we see the powerful picture of Christ's *Imminent Return*.[44] The image itself reveals a dolphin leaping upward out of the water. For the Old Testament dispensation, this decan was a sign of *the Coming Branch*, coming from the waters of the world (Rev. 17:5). For the New Testament dispensation, it is a sign of *the Imminent Return* of Christ which can take place at any moment!

"In light of the Myth"

In ancient Greek mythology, *Apollo*, in the form of a dolphin, showed the Cretans the way to Delphi; the seat of the most famous oracles and the most celebrated place in the Grecian world. Spiritually speaking, one day, in a moment, in the twinkling of an eye, Christ like a dolphin will leap out of heaven and come to snatch his Bride from off the earth. At that moment, the church age concludes and "we shall ever be with the Lord" (1st Thess. 4:17).

---

[44] The Imminent Return is the acknowledgement that the Lord's return is so near that it is constantly on our minds. When one thinks of imminent, he should think of "any moment."

Principal Stars

- *Dalaph* is an Arabic name and means *coming quickly.* "Behold, I come quickly: blessed *is* he that keepeth the sayings of the prophecy of this book." (Revelation 22:7)
- *Scalooin* means *swift.*

In light of the principal stars, Delphinus reminds the believer that his resurrected God is soon to return. Thus, we are to keep our lives living in the constant sphere of God's Word.

The Imminent Coming of the Lord Jesus Christ for his Church was taught:

- By Paul – "For our conversation is in heaven; from whence also we look for the Saviour, the Lord Jesus Christ: Who shall change our vile body, that it may be fashioned like unto his glorious body, according to the working whereby he is able even to subdue all things unto himself" (Phil. 3:20-21). "Let your moderation be known unto all men. The Lord *is* at hand." (Phil. 4:5)
- "For they themselves shew of us what manner of entering in we had unto you, and how ye turned to God from idols to serve the living and true God; And to wait for his Son from heaven, whom he raised from the dead, *even* Jesus, which delivered us from the wrath to come" (1st Thess. 1:9-10).
- By Peter – "Wherefore gird up the loins of your mind, be sober, and hope (ἐλπιδα-expectation) to the end for the grace that is to be brought unto you at the revelation of Jesus Christ; As obedient children, not fashioning yourselves according to the former lusts in your ignorance: But as he which hath called you is holy, so be ye holy in all manner of conversation" (1st Pet. 1:13-15).
- By John – "Beloved, now are we the sons of God, and it doth not yet appear what we shall be: but we know that, when he shall appear, we shall be like him; for we shall see him as he is. And every man that hath this hope (ἐλπιδα-expectation) in him purifieth himself, even as he is pure" (1st Jn. 3:2-3).

These references make an excellent appeal for holy living and diligent service by all Christians. This was to be the attitude of the Old Testament saints as they looked for *the Coming Messiah*. Likewise, it is to be the attitude of the New Testament saints as we look for *the Return of Christ*. We should be waiting, looking, and expecting the blessed appearing of the Savior.

If we are a part of that generation, it may be said of us, "we which are alive *and* remain unto the coming of the Lord shall not prevent them which are asleep. For the Lord himself shall descend from heaven with a shout, with the voice of the archangel, and with the trump of God: and the dead in Christ shall rise first: Then we which are alive *and* remain shall be caught up together with them in the clouds, to meet the Lord in the air: and so shall we ever be with the Lord." (1st Thess. 4:15-17)

What a grand reunion that will be. We will get to see our parents, children, former heroes of the faith, and many friends and family members that have gone on before us. At that moment, we will receive our new and glorified bodies, but most of all—we will get to see Jesus. What a day that will be.

> What a day that will be when my Jesus I shall see,
> And I look upon His face,
> The One who saved me by His grace;
> When He takes me by the hand
> And leads me through the Promised Land,
> What a day, glorious day that will be!
> **(Jim Hill)**

AQUARIUS "THE WATER BEARER"
CHAPTER 6 SECTION 1

In Aquarius, we have an image of a youthful man tipping an urn, from which flows a powerful stream of water. In this image of the invisible God, we see the promised Spirit being poured out of the heavenly vial after the Messiah was "cut off" at Calvary.[45] Aquarius, like the other 11 main constellations, has three decans; *the Southern Fish*, *Pegasus*, and *Cygnus*, all representing an important teaching within the gospel message of Biblical Astrology. This doctrine is the third Person of the Godhead Body, called the Holy Spirit.

The Mythology
In the Greek culture, Aquarius' name is Ganymede, meaning "the bright, glorified and happy one." Ganymede, although an enigma of extravagance on earth, came to an untimely death; upon which, the father of the gods carried him away to heaven on eagle's wings to be with him in glory.

---

[45] Daniel 9:26 "And after threescore and two weeks shall Messiah be cut off, but not for himself: and the people of the prince that shall come shall destroy the city and the sanctuary; and the end thereof *shall be* with a flood, and unto the end of the war desolations are determined."

With the perversion of primeval astrology, we can relate the story to our Messiah, who came into the world as a light and example, and while in his prime was cut off and ascended on high to the right hand of the Father.

The Principal Stars in Aquarius

- *Sheat* is a Hebrew word meaning *who goeth and returneth.* The book of Daniel informs us that the Messiah would come and be cut off (die) at the end of the 69th week of the 70 week prophecy. Daniel 9:26 "And after threescore and two weeks shall Messiah be cut off, but not for himself: and the people of the prince that shall come shall destroy the city and the sanctuary; and the end thereof *shall be* with a flood, and unto the end of the war desolations are determined." Shortly before the "cut off" would come, Jesus promised that he would not leave the church comfortless, but that he would come to them (John 14:18). Christ had to go away to atone for sin, but he came again in the Person of the Holy Spirit on Pentecost. This was fulfilled on the Day of Pentecost and thus, fulfilled the Old Testament prophecy of Joel.

And it shall come to pass afterward, *that* I will pour out my spirit upon all flesh; and your sons and your daughters shall prophesy, your old men shall dream dreams, your young men shall see visions: And also upon the servants and upon the handmaids in those days will I pour out my spirit. And I will show wonders in the heavens and in the earth, blood, and fire, and pillars of smoke. The sun shall be turned into darkness, and the moon into blood, before the great and the terrible day of the LORD come. And it shall come to pass, *that* whosoever shall call on the name of the LORD shall be delivered: for in mount Zion and in Jerusalem shall be deliverance, as the LORD hath said, and in the remnant whom the LORD shall call. (Joel 2:28-32)

- *Mon* or *Meon* an Egyptian word for *"urn."*
- *Ancha* the vessel of *"the pouring out."*
- *Sa'ad al Melik* meaning *"record of the out-pouring."*

This Old Testament prophecy of the pouring out of the Spirit is said to have been fulfilled in the New Testament in Acts.

But this is that which was spoken by the prophet Joel; And it shall come to pass in the last days, saith God, I will pour out of my Spirit upon all flesh: and your sons and your daughters shall prophesy, and your young men shall see visions, and your old men shall dream dreams: And on my servants and on my handmaidens I will pour out in those days of my Spirit; and they shall prophesy: And I will show wonders in heaven above, and signs in the earth beneath; blood, and fire, and vapor of smoke: The sun shall be turned into darkness, and the moon into blood, before that great and notable day of the Lord come: And it shall come to pass, *that* whosoever shall call on the name of the Lord shall be saved. (Acts 2:16-21)

Spiritual and Literal Baptism

While Peter yet spake these words, the Holy Ghost fell on all them which heard the word. And they of the circumcision which believed were astonished, as many as came with Peter, because that on the Gentiles also was poured out the gift of the Holy Ghost. For they heard them speak with tongues, and magnify God. Then answered Peter, Can any man forbid water, that these should not be baptized, which have received the Holy Ghost as well as we? And he commanded them to be baptized in the name of the Lord. Then prayed they him to tarry certain days. (Acts 10:44-48)

This passage is important in that it shows both, the spiritual baptism of the pouring out of the Holy Ghost and the physical baptism ("can any man forbid water"). It also shows that the physical is a representation of the spiritual. The Spiritual is the assurance of one's salvation, and the physical is an outward showing of the inward profession. In the passage below, Peter shows that the inward baptism is the indwelling of the Holy Spirit.

And as I began to speak, the Holy Ghost fell on them, as on us at the beginning. Then remembered I the word of the Lord, how that he said, John indeed baptized with water; but ye shall be baptized with the Holy Ghost. (Acts 11:15-16)

And when there had been much disputing, Peter rose up, and said unto them, Men *and* brethren, ye know how that a good while ago God made choice among us, that the Gentiles by my mouth should hear the word of the gospel, and believe. And God, which knoweth the hearts, bare them witness, giving them the Holy Ghost, even as *he* did unto us; And put no difference between us and them, purifying their hearts by faith. (Act 15:7-9)

History of the Jewish Baptism

Sa'ad al Su'ud means *the flowing stream*. The Hebrew word for baptism is pronounced in English as *meek-vah*. It originated with the priests as a ceremonial washing before going into the Temple and later was converted into baptism. The ancient method of Baptism was done in moving water such as a spring or river, at about chest high. The person being baptized would go completely under the water upon his own volition three times. There had to be at least one witness, namely the baptizer. This individual explained all the religious rites and procedures of baptism. John "the forerunner or messenger" (Mal. 3:1; Matt. 3:1-3; Mark 1:2) was a baptizer. Matthew 3:6 says that many were "baptized of him in Jordan." The preposition *of* is represented from by underlying Greek word υπο, pronounced in English as *hoo-po*. *Hoo-po* means that they were baptized **under** John's supervision. John didn't grab their nose and take them under the water, but they were taken under the water by themselves under John's supervision. This would have been the baptism that Jesus Christ understood and was baptized with. John forbade to baptize Jesus because John understood that Jesus was the Author of baptism, and John was unqualified to baptize him knowing that he was a sinner, and Christ was sinless. John's acknowledgment of these things gave a beautiful declaration and fulfillment to Christ's Messiah-ship.

"Then cometh Jesus from Galilee to Jordan unto John, to be baptized of (υπο) him. But John forbade him, saying, I have need to be baptized of thee, and comest thou to me? And Jesus answering said unto him, Suffer *it to be so* now: for thus it becometh us to fulfill all righteousness. Then he suffered him. And Jesus, when he was baptized, went up straightway out of the water: and, lo, the heavens were opened unto him, and he saw the Spirit of God descending like a dove, and lighting upon him: And lo a

76

voice from heaven, saying, This is my beloved Son, in whom I am well pleased." (Matt. 3:13-17)

In Aquarius, an image of the invisible God, we see the promised Spirit being poured out of the heavenly vial after the Messiah was "cut off" at Calvary. The Holy Spirit is our assurance of Salvation and is symbolized as *a stream of water,* not a stagnant pool. Significantly, Baptism became the first ordinance of the Early Church. The church demonstrated this outward showing (immersion under water three times) signifying that the individual had acknowledged Christ as the Messiah, the Father as Christ's Sender, and the Holy Spirit as Christ's Abider.[46]

> For I will pour water upon him that is thirsty, and floods upon the dry ground: I will pour my spirit upon thy seed, and my blessing upon thine offspring. (Isaiah 44:3)

> And the Spirit and the bride say, Come. And let him that heareth say, Come. And let him that is athirst come. And whosoever will, let him take the water of life freely. (Revelation 22:17)

---

[46]Go ye therefore, and teach all nations, baptizing them in the name of the Father, and of the Son, and of the Holy Ghost: (Matt. 28:19)

Fom al Haut

As we have already seen, and will see multiple times in the astrological record, Christians are represented as fish. In this image, we see a fish swimming and drinking from the river of life issuing out of the urn which Aquarius has poured forth. Pisces Australis has one principal star, named Fom al Haut.

PRINCIPAL STAR
- *Fom al Haut*- meaning, "the mouth of the fish."

IN LIGHT OF THE MYTHS
In the myth's perversion of the original astrological message, we find the story of Astarte, known by the Greeks as Aphrodite. After being attacked and trying to escape Typhon (the enemy), she metamorphosed herself into a fish.

SEEING THE TRUTH THROUGH THE MYTH
Likewise, in order for Christians to escape the power of Satan and his hold on the world, man must be metamorphosed into a new creature. Paul said to the church at Corinth, "therefore if any man *be* in Christ, *he is* a new creature: old things are passed away; behold, all things are become new" (II Cor. 5:17). This constellation represents the Christian enjoying the benefits of his new life in Christ; a life of faith, trust and continual belief as he drinks from the spiritual fountain that never runs dry. Paul pleaded

with the church at Galatia saying, "My little children, of whom I travail in birth again until Christ be formed (μορφωθῇ - morphed) in you" (Gal. 4:19). Paul also told the Church at Rome that they needed a metamorphosis. "I beseech you therefore, brethren, by the mercies of God, that ye present your bodies a living sacrifice, holy, acceptable unto God, *which is* your reasonable service. And be not conformed to this world: but be ye transformed (μεταμορφοῦσθε - metamorphosis) by the renewing of your mind, that ye may prove what *is* that good, and acceptable, and perfect, will of God" (Rom. 12:1-2).

After salvation, the Devil knows that he needs an inroad in order to reach the Christian, so he capitalizes on those who refuse to apply the Word of God to their lives. Christ wants to be *formed* (morphed) in us; therefore, by doing his will, we allow him to live through us. It is then that we become vessels for his use; and after taking control of our lives, we begin to see "Christ in us the hope of glory" (par. Col. 1:27). The image of the Southern Fish is portrayed with its mouth agape and drinking from the eternal waters of Aquarius. In John 4:13-14, Jesus said to the woman at the well, "Whosoever drinketh of this water shall thirst again: But whosoever drinketh of the water that I shall give him shall never thirst; but the water that I shall give him shall be in him a well of water springing up into everlasting life." Dear reader, are you drinking from the waters that never run dry?

"Dwelling in Beulah Land"

Far away the noise of strife upon my ear is falling,
Then I know the sins of earth beset on every hand;
Doubt and fear and things of earth in vain to me are calling,
None of these shall move me from Beulah Land.

Chorus: I'm living on the mountain, underneath a cloudless sky,
I'm drinking at the fountain that never shall run dry;
O yes, I'm feasting on the manna from a bountiful supply,
For I am dwelling in Beulah Land.

Far below the storm of doubt upon the world is beating,
Sons of men in battle long the enemy withstand;
Safe am I within the castle of God's word retreating,
Nothing then can reach me, 'tis Beulah Land. (Chorus)

Let the stormy breezes blow, their cry cannot alarm me,
I am safely sheltered here, protected by God's hand;
Here the sun is always shining, here there's naught can harm me,
I am safe forever in Beulah Land. (Chorus)

Viewing here the works of God, I sink in contemplation,
Hearing now His blessed voice, I see the way is planned;
Dwelling in the spirit, here I learn of full salvation,
Gladly will I tarry in Beulah Land. (Chorus)

(Charles Austin Miles 1868-1946)

PEGASUS (THE WINGED HORSE)
CHAPTER 6 SECTION 3

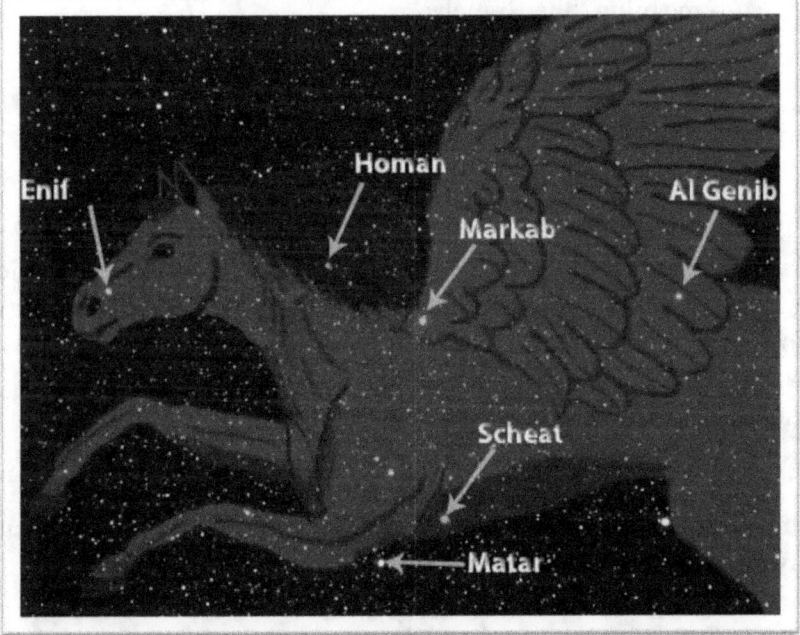

"Sing unto God, sing praises to his name: extol him that rideth upon the heavens by his name JAH, and rejoice before him." (Psalm 68:4) Pegasus is the second decan in the house of Aquarius, and means "horse of the gushing fountain."

IN LIGHT OF THE MYTHS
The myths inform us that this horse arose into being from the slaying of Medusa by Perseus. Thus it was named Pegasus, *the horse of the fountain*. His name is derived because he first appeared near the springs of a mighty sea. He is said to have lived in the palace of the great king and father of gods. He thundered and lightened for Jupiter, until Bellephron gained possession of him through sacrifice to the goddess of justice. Followed by a deep sleep, he was given a golden bridle in which the wild horse obeyed. Although he was brought forth to victory, it was not without receiving a painful sting in his foot, at which time Pegasus took his wings.

- *Markab*- meaning "the returning."
- *Scheat*- meaning "he who goeth and returneth."
- *Enif*- meaning "the Branch."
- *Al Genib*- meaning "who carries."
- *Homan*- meaning "the waters."
- *Matar*- meaning "who causeth the plenteous to overflow."

SEEING THE TRUTH BEHIND THE MYTH

This sign, although obscured by the myths, is a plain declaration of *the Coming Branch* prophesies and another proto-evangelism of Genesis 3:15's gospel. Jesus came into the world (signified by the water) as "the Coming Branch." He was stung in the foot (Genesis 3:15) on Calvary by which he entered into a deep sleep and came up out of the grave victorious. "And I will put enmity between thee and the woman, and between thy seed and her seed; it shall bruise thy head, and thou shalt bruise his heel." (Genesis 3:15) "In that day shall the branch of the LORD be beautiful and glorious, and the fruit of the earth *shall be* excellent and comely for them that are escaped of Israel" (Isaiah 4:2). "And there shall come forth a rod out of the stem of Jesse, and a Branch shall grow out of his roots" (Isaiah 11:1). "In those days, and at that time, will I cause the Branch of righteousness to grow up unto David; and he shall execute judgment and righteousness in the land." (Jeremiah 33:15) "Hear now, O Joshua the high priest, thou, and thy fellows that sit before thee: for they *are* men wondered at: for, behold, I will bring forth my servant the BRANCH." (Zechariah 3:8) "And speak unto him, saying, Thus speaketh the LORD of hosts, saying, Behold the man whose name *is* The BRANCH; and he shall grow up out of his place, and he shall build the temple of the LORD." (Zechariah 6:12)

CYGNUS (THE SWAN)
CHAPTER 6 SECTION 4

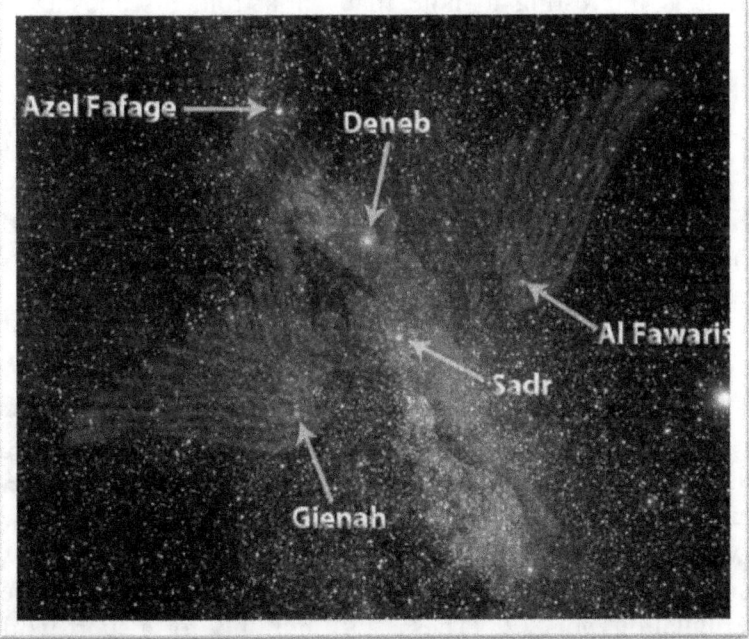

According to Seiss, the swan is "on the wing, in the act of rapid flight, circling and returning, as its names in Greek and Latin signify."[47] The principal stars which mark the height and breadth of the swan form what many modern-day astronomers call the "Northern Cross."

PRINCIPAL STARS

- *Deneb*- meaning "the Judge to come."
- *Adige*- same as Deneb meaning "flying swiftly."
- *Arided*- same as Deneb meaning "he shall come down."
- *Azel-Fafage*—Azel meaning "who goes and returns" and Fafage meaning "glorious shining forth."
- *Sadr*- meaning "who returns as in a circle."

The Cygnus signifies the rapid coming of the Lord Jesus to the earth in his First Advent. In this image, we see *the First Advent* of Jesus Christ and his promised Holy Spirit's return. Cygnus was a reminder to the world of

---

[47] Seiss, Joseph, *the Gospel in the Stars*, The Castle Press, Philadelphia, 1884

the coming Messiah. When thinking of the stars *Azel,* meaning "who goes and returns," and *Sadr,* meaning "who returns as in a circle," we see reference to another coming of this Earthly Bird. With the previous stars in mind, it is interesting to note that Christ told his apostles in John, "Nevertheless I tell you the truth; It is expedient for you that I go away: for if I go not away, the Comforter will not come unto you; but if I depart, I will send him unto you. And when he is come, he will reprove the world of sin, and of righteousness, and of judgment: Of sin, because they believe not on me; Of righteousness, because I go to my Father, and ye see me no more; Of judgment, because the prince of this world is judged." (John 16:7-11). Again, Christ said in John 14:3, "and if I go and prepare a place for you, I will come again, and receive you unto myself; that where I am, *there* ye may be also."

It is this Swan that works ever so powerfully to do the work of God and build his Church. It was his Spirit that swept through the heavens to save our souls from the pits of the world and the flames of Hell. One day, at the rapture of the saints, the Swan will be taken up from off this world and ascend with the Church (2nd Thess. 2:6-8). After this event, with the restraining influence of the Holy Spirit gone, the Antichrist will be free to do his Tribulation work (2nd Thess. 2:9), but only for a short while (Rev. 12:12).

PISCES (THE DISPENSATIONAL FISHES)
CHAPTER 7 SECTION 1

Pisces, like the other 11 main constellations, has three decans; *the Band*, *Cepheus*, and *Andromeda*, and presents a unique dispensational truth in *Biblical Astrology's gospel*. In this image, we see two fish. These fish are not only two separate and distinct fish, but represent two separate and distinct groups; namely, the nation of Israel and the Church. These fish are bound by bands and tied around the neck of Cetus. The heavenly image teaches that the promises of God's covenant blessings are only partial **until the bands are broken**. The presence of the enemy has constantly kept the nation of Israel and the Church in check. Likewise, the presence of the Church checks the enemy. This is signified by the bands being tied to the neck of Cetus. The Bible informs us in Romans that "the earnest expectation of the creature waiteth for the manifestation of the sons of God. For the creature was made subject to vanity, not willingly, but by reason of him who hath subjected *the same* in hope, Because the creature itself also shall be delivered from the bondage of corruption into the glorious liberty of the children of God. For we know that the whole creation groaneth and travaileth in pain together until now." (Romans 8:19-22). In these verses, we see that the creation is subject to vanity.

- *Al Samaca* - meaning *"the upheld."*
- *Okda* - meaning *"the united."*

Okda is significant because in Heaven we will not be divided by the dispensational cultural context, but universally connected in one universal brotherhood. This representation is seen by the 24 elders seated around the throne (Rev. 4:4-10; 5:5-14; 7:11-13; 11:16; 14:3; 19:4), as well as in the heavenly city's wall having twelve gates with the twelve tribes of Israel written in them (Rev. 21:12), with the wall's foundation containing the names of the twelve apostles (Rev. 21:14). The dispensational view of the Church was no doubt a mystery to the nation of Israel; therefore, what we are capable of seeing from our age is ideally different from what the dispensation in the Law understood.

The Mystery Revealed

Now to him that is of power to establish you according to my gospel, and the preaching of Jesus Christ, according to the revelation of the mystery, which was kept secret since the world began. But now is made manifest, and by the Scriptures of the prophets, according to the commandment of the everlasting God, made known to all nations for the obedience of faith. (Rom. 16:25-26)

Even the mystery which hath been hid from ages and from generations, but now is made manifest to his saints: To whom God would make known what *is* the riches of the glory of this mystery among the Gentiles; which is Christ in you, the hope of glory. (Col. 1:26-27)

If ye have heard of the dispensation of the grace of God which is given me to you-ward: How that by revelation he made known unto me the mystery; (as I wrote afore in few words, Whereby, when ye read, ye may understand my knowledge in the mystery of Christ) Which in other ages was not made known unto the sons of men, as it is now revealed unto his holy apostles and prophets by the Spirit; That the Gentiles should be fellow heirs, and of the same body, and partakers of his promise in Christ by the gospel: Whereof I was made a minister, according to the gift of the grace of God given unto me by the effectual working of his power. Unto me, who am less than the least of all saints, is this grace given, that I should preach among the Gentiles the unsearchable riches of Christ; And to make all *men* see what *is* the fellowship of the mystery, which from the beginning of the world hath been hid in God, who created all things by Jesus Christ: To the intent that now unto the principalities and powers in heavenly *places* might be known by the church the manifold wisdom of God, According to the eternal purpose which he purposed in Christ Jesus our Lord: In whom we have boldness and access with confidence by the faith of him. (Eph. 3:2-12)

The stars in Pisces are still speaking to us today. They are reminding us that someday all evil will be suppressed, and God will no longer be known through a dispensational cloud. Mankind will know him for who he is and in all of his glory. What a great eternal day that will be. "Beloved, now are we the sons of God, and it doth not yet appear what we shall be: but we know that, when he shall appear, we shall be like him; for we shall see him as he is. And every man that hath this hope in him purifieth himself, even as he is pure." (1st John 3:2-3)

THE BAND
CHAPTER 7 SECTION 2

In the heavenly image, the two fishes of Pisces are bound by *the Band*. In most images, *the Band* has the tails of two fishes, which are fastened to the neck of Cetus. The same image seen in *the banded fishes* is also seen in the decan of Andromeda, *the chained woman*. In spite of these images of bondage telling one in the same story, it is significant to note that the Deliverer from bondage is ever near in Cepheus (see star chart in the back of the book).

SIX BANDS FOR GOD'S PEOPLE
It is worth knowing that there are six noteworthy bands for God's people:

1) Time- The band of time is a constant reminder of our limited existence. We can never have enough time in the day to get all the work done. Time is a necessary band, which taunts the human psyche to ponder everlasting life. With a limited reality facing us every day, we can better understand the eternality of Heaven. Once a person becomes saved through the sacrifice of Christ, he can

receive peace of mind about his eternal condition by desiring, longing for, and expecting Heaven.

2) Finite- As humans, we are completely bound with limitations: physically, mentally, and spiritually. God has a schedule and an architectural and sovereign blueprint to everything, so even when our spirituality is strong, God may weaken or limit it through waiting, suffering and persecution, in order to accomplish his higher plan.

3) Flesh- Our carnality is a band, in that it limits our understanding of the infinite. It makes us trust, believe, hope and place our faith in the invisible God.

4) Sin- Our sinful condition is a constant band that wears at us and makes us realize our frailty and weaknesses. However, it is through this band that we are able to see the grace, mercy and love of God.

5) Satan- The band of the Satan is always there. In spite of all the natural and supernatural bands, there is a band by Satan himself to thwart and divert us from the plan of God. He is "as a roaring lion," walking about, "seeking whom he may devour" (1st Pet. 5:8).

6) Progressive Revelation- Finally, God has only revealed himself throughout the millennia by piecemeal (Heb 1:1). With man's limited revelation of God, there have been things hidden from the foundation of the world.

7)

PRINCIPAL STAR

- *Al Risha* - meaning "the band." These appear to be the same bindings that hold Andromeda. Cicero called them Vincla which interpreted is "the bonds."[48]

In spite of Israel and the Church's bands, we can still have foretastes of divine glory through our bands. Much of Christianity's glory is experiencing as many of these celestial insights as possible. We can easily see that in Christ man can have victory over all of his bands. "Beloved, now are we the sons of God, and it doth not yet appear what we shall be: but we know that, when he shall appear, we shall be like him; for we shall see him as he is. And every man that hath this hope in him purifieth himself, even as he is pure." (1st Jn. 3:2-3) "For now we see through a glass, darkly; but then face to face: now I know in part; but then shall I know even as also I am known." (1st Corinthians 13:12)

---

[48] Marcus Tullius Cicero, 106-43 B.C., orator and versifier of Aratos, 3, 272; *et passim.*

"What a Day that Will Be" (Jim Hill 1955)

There is coming a day,
When no heart aches shall come;
No more clouds in the sky,
No more tears to dim the eye;
All is peace forever more,
On that happy golden shore;
What a day, glorious day that will be.

Chorus: What a day that will be,
When my Jesus I shall see;
And I look upon his face,
The One who saved me by His grace;
When He takes me by the hand,
And leads me through the Promised Land;
What a day, glorious day that will be.

There'll be no sorrow there,
No more burdens to bear;
No more sickness, and no more pain,
And no more parting over there;
And forever I will be,
With the One who died for me;
What a day, glorious day that will be.

Chorus

CEPHEUS "THE KING"
CHAPTER 7 SECTION 3

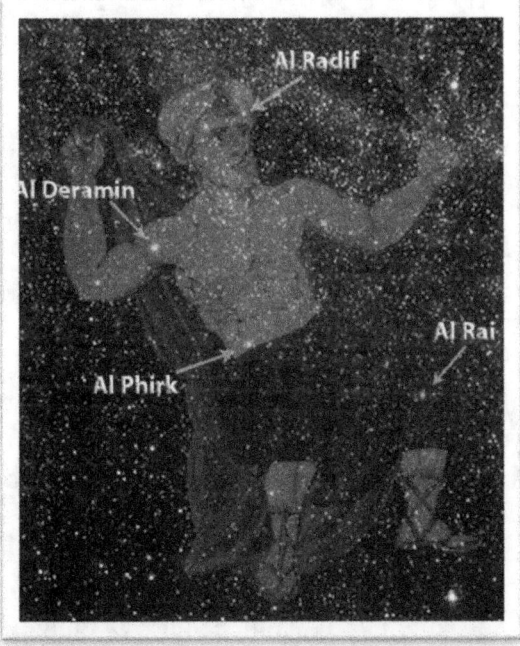

"Which he wrought in Christ, when he raised him from the dead, and set *him* at his own right hand in the heavenly *places,*   Far above all principality, and power, and might, and dominion, and every name that is named, not only in this world, but also in that which is to come: And hath put all *things* under his feet, and gave him *to be* the head over all *things* to the church, Which is his body, the fulness of him that filleth all in all." (Eph. 1:20-23)

Cepheus is the third decan of Pisces and powerfully pictures Christ as the Universal King. Of all the images in the heavens, this is the one that gives us Christ as the crowned King. In the Greek, Cepheus means the Royal Branch, which is the same Branch that Virgo holds. This constellation is unique in that it sits unopposed above the universe, being positioned in the heavens without reacting to any other constellation. Thus, we find the exalted attribute of God's sovereignty and Christ's Kingship.

- Al Deramin- Arabic name meaning, *"the quickly returning or coming quickly."* Al Deramin is the brightest star located in the right shoulder.
- Al Phirk- Arabic name meaning, *"the Redeemer."* This star is located in Cepheus' girdle.
- Al Rai- Arabic name meaning, *"the shepherd."*
- Regulus- meaning, *"treading under foot."*

These stars further depict the King whose name is the Branch, as seen in the Old Testament's Apocalypse, Zechariah. "And speak unto him, saying, Thus speaketh the LORD of hosts, saying, Behold the man whose name *is* The BRANCH; and he shall grow up out of his place, and he shall build the temple of the LORD: Even he shall build the temple of the LORD; and he shall bear the glory, and shall sit and rule upon his throne; and he shall be a priest upon his throne: and the counsel of peace shall be between them both" (6:12-13). This King came unto Zion, but was rejected by divine providence and for the sole purpose of a greater Gentile kingdom.

"Rejoice greatly, O daughter of Zion; shout, O daughter of Jerusalem: behold, thy King cometh unto thee: he *is* just, and having salvation; lowly, and riding upon an ass, and upon a colt the foal of an ass" (Zech. 9:9). "All this was done, that it might be fulfilled which was spoken by the prophet, saying, Tell ye the daughter of Sion, Behold, thy King cometh unto thee, meek, and sitting upon an ass, and a colt the foal of an ass." (Matt. 21:4-5)

When God's time is come, Christ will be high and lifted up "and the LORD shall be king over all the earth: in that day shall there be one LORD, and his name one" (Zech. 14:9). This verse parallels perfectly with Cepheus' foot on the North Star; which is the central star of the Northern Planisphere, indicating that this King is the Universal King. He is the Sovereign and the Omnipotent. He is the King of kings and the Lord of lords. "And it shall come to pass, *that* every one that is left of all the nations which came against Jerusalem shall even go up from year to year to worship the King, the LORD of hosts, and to keep the feast of tabernacles." (Zech. 14:16).

Man has yet to know this part of Christ. We can only have a foretaste on this side of eternity, as we are constantly struggling with our sin nature to let him rule in our lives. By Adam's own free will, God was dethroned from man's heart. Because of our free moral choice in receiving Christ as Savior, we will someday see "the Lord sitting upon a throne, high and lifted up" with his train filling the temple (Is. 6:1). We will know him in our new and glorified bodies, without the sin nature.

The stars in Cepheus are still speaking to us today. They are reminding us that one day and without sin, we will see and know Christ as King in all his glory, power, and majesty. We "shall know even as we are known." As Cepheus in the night sky, so Christ in Heaven's future will forever be unopposed. This image is a magnificent foreshadow of Jesus Christ's divine and royal glory. And we ask the question, "Why must we die?"!

"Jesus Shall Reign"

Jesus shall reign where'er the sun
Does his successive journeys run;
His kingdom stretch from shore to shore,
Till moons shall wax and wane no more.

Behold the islands with their kings,
And Europe her best tribute brings;
From north to south the princes meet,
To pay their homage at His feet.

There Persia, glorious to behold,
There India shines in eastern gold;
And barb'rous nations at His word
Submit, and bow, and own their Lord.

To Him shall endless prayer be made,
And praises throng to crown His head;
His Name like sweet perfume shall rise
With every morning sacrifice.

People and realms of every tongue
Dwell on His love with sweetest song;
And infant voices shall proclaim
Their early blessings on His Name.

Blessings abound wherever He reigns;
The prisoner leaps to lose his chains;
The weary find eternal rest,
And all the sons of want are blessed.

Where He displays His healing power,
Death and the curse are known no more:
In Him the tribes of Adam boast
More blessings than their father lost.

Let every creature rise and bring
Peculiar honors to our King;
Angels descend with songs again,
And earth repeat the loud amen!

Great God, whose universal sway
The known and unknown worlds obey,
Now give the kingdom to Thy Son,
Extend His power, exalt His throne.

The scepter well becomes His hands;
All Heav'n submits to His commands;
His justice shall avenge the poor,
And pride and rage prevail no more.

With power He vindicates the just,
And treads th'oppressor in the dust:
His worship and His fear shall last
Till hours, and years, and time be past.

As rain on meadows newly mown,
So shall He send his influence down:
His grace on fainting souls distills,
Like heav'nly dew on thirsty hills.

The heathen lands, that lie beneath
The shades of overspreading death,
Revive at His first dawning light;
And deserts blossom at the sight.
The saints shall flourish in His days,
Dressed in the robes of joy and praise;
Peace, like a river, from His throne
Shall flow to nations yet unknown.

Isaac Watts (1719)

Psalm 72

Give the king thy judgments, O God, and thy righteousness unto the king's son. He shall judge thy people with righteousness, and thy poor with judgment. The mountains shall bring peace to the people, and the little hills, by righteousness. He shall judge the poor of the people, he shall save the children of the needy, and shall break in pieces the oppressor. They shall fear thee as long as the sun and moon endure, throughout all generations. He shall come down like rain upon the mown grass: as showers *that* water the earth. In his days shall the righteous flourish; and abundance of peace so long as the moon endureth. He shall have dominion also from sea to sea, and from the river unto the ends of the earth. They that dwell in the wilderness shall bow before him; and his enemies shall lick the dust. The kings of Tarshish and of the isles shall bring presents: the kings of Sheba and Seba shall offer gifts. Yea, all kings shall fall down before him: all nations shall serve him. For he shall deliver the needy when he crieth; the poor also, and *him* that hath no helper. He shall spare the poor and needy, and shall save the souls of the needy. He shall redeem their soul from deceit and violence: and precious shall their blood be in his sight. And he shall live, and to him shall be given of the gold of Sheba: prayer also shall be made for him continually; *and* daily shall he be praised. There shall be an handful of corn in the earth upon the top of the mountains; the fruit thereof shall shake like Lebanon: and *they* of the city shall flourish like grass of the earth. His name shall endure for ever: his name shall be continued as long as the sun: and *men* shall be blessed in him: all nations shall call him blessed. Blessed *be* the LORD God, the God of Israel, who only doeth wondrous things. And blessed *be* his glorious name for ever: and let the whole earth be filled *with* his glory; Amen, and Amen. The prayers of David the son of Jesse are ended.

THE ANDROMEDA
(LIBERATED FROM CAPTIVITY)
CHAPTER 7 SECTION 4

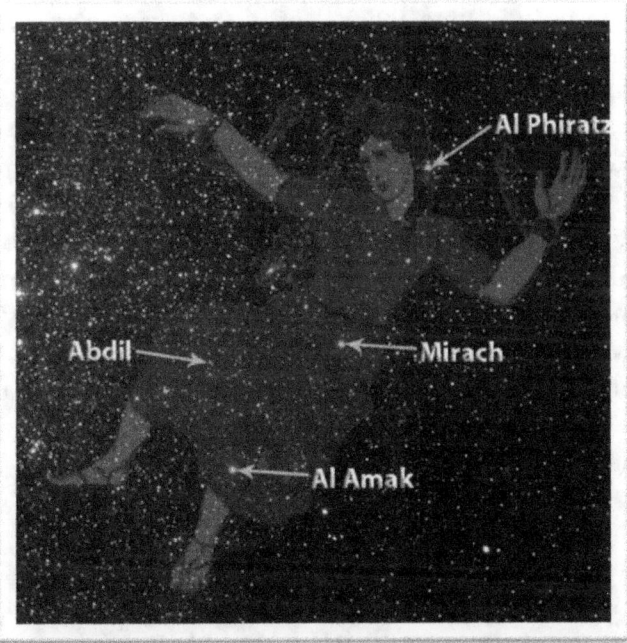

In Pisces, we saw symbolically *the band* which restricted the Church of God and the nation of Israel. In Andromeda, we see the **universal** Bride of God liberated from her shackles and gloriously delivered by her King. The image of Andromeda is one of bondage, but the story of Andromeda, in light of her sister decans is one of liberation from bondage. The bondage that she is delivered from is given in the bands of Pisces. Therefore, Pisces, the Bands, and Andromeda are one in the same story.

PRINCIPAL STARS OF ANDROMEDA

- *Abdil* meaning "afflicted."
- *Mirach* meaning "the weak."
- *Al Mara* meaning "the afflicted."
- *Persea* meaning "the stretched out."
- *Al Moselsalah* meaning "from the grave."
- *Al Phiratz* meaning "the broken down."
- *Al Amak* meaning "the struck down."
- *Misam al Thuraiya* meaning "the assembled."

97

- *Desma* meaning "the bound."
- *Andromedia* meaning "the set free, or liberated."

The Universal Church of God

In Heaven, we will become the one true and Universal Church. Matthew 19:28 "And Jesus said unto them, Verily I say unto you, That ye which have followed me, in the regeneration when the Son of man shall sit in the throne of his glory, ye also shall sit upon twelve thrones, judging the twelve tribes of Israel." What is unique in Christ's teaching here is that the church will be incorporated with the nation of Israel in a universal brotherhood. This will only be after God's plan for *the Church Age* ends. Over and over again in the stars, we see the same story of the Bible. It was revealed in pieces throughout the Old Testament and unveiled in the New Testament. "Do ye not know that the saints shall judge the world? and if the world shall be judged by you, are ye unworthy to judge the smallest matters?" (1st Corinthians 6:2) "And hath made us kings and priests unto God and his Father; to him be glory and dominion for ever and ever. Amen." (Revelation 1:6)

ARIES (THE EXALTED LAMB)
CHAPTER 8 SECTION 1

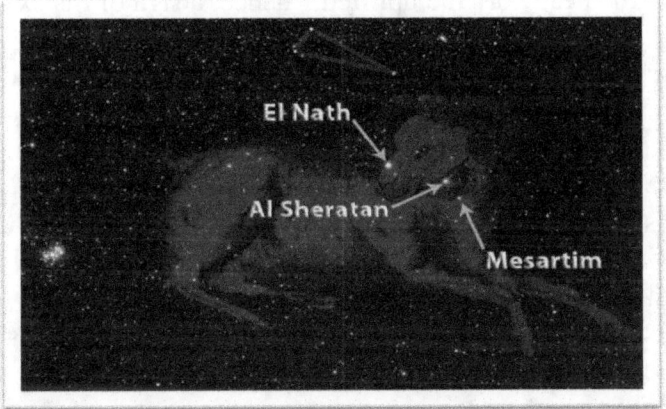

Modern images of Aries show an image of a Ram, but its more ancient signs show that of a Lamb. In Syriac, Aries is pronounced in English as *Amroo* and means "the Lamb." The Arab world calls the name of this sign *Al Hamal,* meaning *"the Sheep, the Gentle,* or *the Merciful."* In the ancient **Zodiac of Denderah** in Egypt, the sign of Aries has no horns and is crowned with a halo. There is no contradiction whether the Lamb or the Ram is signified because they are one in the same picturing Christ. This is seen in the story of Abraham and Isaac below. Aries, like the other 11 main constellations, has three decans; *Cassiopeia, Cetus,* and *Perseus,* which teach another part of the gospel story in *Biblical Astrology.*

> And Isaac spoke unto Abraham his father, and said, My father: and he said, Here *am* I, my son. And he said, Behold the fire and the wood: but where *is* **the lamb** for a burnt offering? And Abraham said, My son, God will provide himself **a lamb** for a burnt offering: so they went both of them together. And Abraham lifted up his eyes, and looked, and behold behind *him* **a ram** caught in a thicket by his horns: and Abraham went and took **the ram**, and offered him up for a burnt offering in the stead of his son. (Gen. 22:7-8; 13)

Aries is the eighth sign in the Zodiac along the sun's path. The name Aries is called *the Chief, the Head* and *Lordly.* Aries is another heavenly image of Christ, who is the Chief, Head, and Lord of his Church.

THE TRIANGLE

Over the head of Aries, there is a magnificent triangle that the ancient Greeks attributed to Deity. The principal star in the triangle bears a name that means "the head" or "the uplifted," hence, a sign of his exaltation.

- Mark 16:19, "So then after the Lord had spoken unto them, he was received up into heaven, and sat on the right hand of God."
- Luke 22:69, "Hereafter shall the Son of man sit on the right hand of the power of God."
- Acts 2:33, "Therefore being by the right hand of God exalted, and having received of the Father the promise of the Holy Ghost, he hath shed forth this, which ye now see and hear."
- Acts 5:31, "Him hath God exalted with his right hand *to be* a Prince and a Savior, for to give repentance to Israel, and forgiveness of sins."
- Acts 7:55, "But he, being full of the Holy Ghost, looked up steadfastly into heaven, and saw the glory of God, and Jesus standing on the right hand of God."
- Acts 7:56, "And said, Behold, I see the heavens opened, and the Son of man standing on the right hand of God."
- Romans 8:34, "Who *is* he that condemneth? *It is* Christ that died, yea rather, that is risen again, who is even at the right hand of God, who also maketh intercession for us."
- 2nd Corinthians 6:7, "By the word of truth, by the power of God, by the armor of righteousness on the right hand and on the left."
- Colossians 3:1, "If ye then be risen with Christ, seek those things which are above, where Christ sitteth on the right hand of God."
- Hebrews 10:12, "But this man, after he had offered one sacrifice for sins forever, sat down on the right hand of God."
- Hebrews 12:2, "Looking unto Jesus the author and finisher of *our* faith; who for the joy that was set before him endured the cross, despising the shame, and is set down at the right hand of the throne of God."
- 1st Peter 3:22, "Who is gone into heaven, and is on the right hand of God; angels and authorities and powers being made subject unto him."

Principal Stars

- *El Nath or El Natik*, meaning "wounded and slain."
- *Al Sheratan*, meaning "the bruised" and "the wounded."
- *Mesartim*, (Hebrew) meaning "the bound."

THE DEBASED LAMB

These stars picture the Lamb that John the Baptist introduced as one, "which taketh away the sin of the world" (John 1:29). John, while on the isle of Patmos, also envisioned Christ as "the Lamb slain from the foundation of the world" (Revelation 13:8). Isaiah prophesied that "he is brought as **a lamb to the slaughter**, and as **a sheep** before her shearers is dumb, so he openeth not his mouth." (Isaiah 53:5-7)

THE EXALTED LAMB

"Saying with a loud voice, Worthy is the Lamb that was slain to receive power, and riches, and wisdom, and strength, and honor, and glory, and blessing" (Revelation 5:12). (See also Rev. 5:6, 8, 13; 6:1, 16; 7:9, 10, 14, 17; 12:11; 13:8; 14:1, 4, 10; 15:3; 17:14; 19:7, 9; 21:9, 14, 22, 23, 27; and 22:1, 3.) Everyone who wishes to know the exalted Lamb should first know the debased Lamb. This is the Lamb who left Heaven to come to this earth and die for our sins. Because of Christ's willingness and work at Calvary, he is exalted "far above all principality, and power, and might, and dominion, and every name that is named, not only in this world, but also in that which is to come" (Eph. 1:21).

CASSIOPEIA (THE EXALTED BRIDE)
CHAPTER 8 SECTION 2

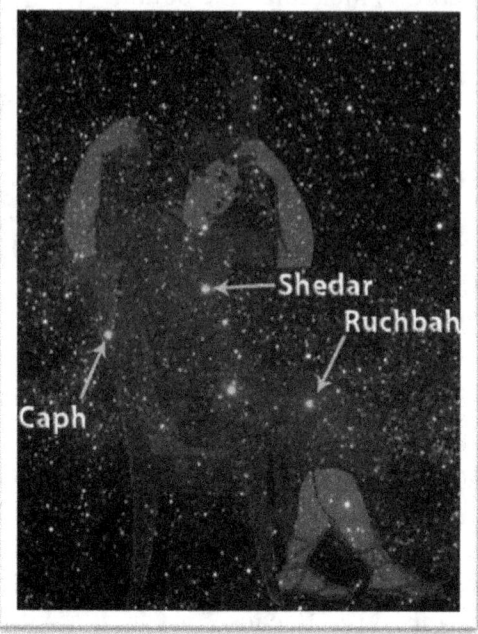

Cassiopeia is the first decan in the Constellation of Aries. Its image is an enthroned woman. Ulugh Beg said that the Arabic name for Cassiopeia is *El Seder,* meaning *"the freed."* In the Denderah Zodiac, her name is Set, meaning "set up as Queen."

Principal Stars

- *Cassiopeia* means "the enthroned."
- *Shedar* means "the freed."
- *Caph* means "the branch."
- *Ruchbah* means "the enthroned."
- *Dat al Cursa* meaning "the seated."

Her story is none other than the inter-dispensational and universal church of God from all ages. In Andromeda, we saw the Church in bondage, soon to be liberated or freed. Here in Cassiopeia, she is delivered from the bands in Pisces, and exalted as the bride of Cepheus. Being rightly positioned with Cepheus in the heavens, she resembles both Israel and Spiritual Israel (the New Testament Church of God); Israel, in that she has

102

and carry's the Branch (*Caph*—the coming Messiah), and the Church, in that she was grafted into Israel. This engrafting is taught in Romans 11:26 as "all Israel."

> For if thou wert cut out of the olive tree which is wild by nature, and wert grafted contrary to nature into a good olive tree: how much more shall these, which be the natural *branches,* be grafted into their own olive tree? For I would not, brethren, that ye should be ignorant of this mystery, lest ye should be wise in your own conceits; that blindness in part is happened to Israel, until the fullness of the Gentiles be come in. And **so all Israel shall be saved**: as it is written, There shall come out of Zion the Deliverer, and shall turn away ungodliness from Jacob: For this *is* my covenant unto them, when I shall take away their sins. As concerning the gospel, *they are* enemies for your sakes: but as touching the election, *they are* beloved for the fathers' sakes. For the gifts and calling of God *are* without repentance. For as ye in times past have not believed God, yet have now obtained mercy through their unbelief: Even so have these also now not believed, that through your mercy they also may obtain mercy. For God hath concluded them all in unbelief, that he might have mercy upon all. O the depth of the riches both of the wisdom and knowledge of God! how unsearchable *are* his judgments, and his ways past finding out! (Romans 11:24-33)

The stars of Cassiopeia are still speaking to us today. They are telling us that one day we will be positioned with our King in the heavens. In that time, there will be no differences in the dispensational congregations, but a single and united Bride culminating into one great and universal Messianic Kingdom.

Romans 2:29, "But he *is* a Jew, which is one inwardly; and circumcision *is that* of the heart, in the spirit, *and* not in the letter; whose praise *is* not of men, but of God."

CETUS (THE SEA MONSTER)
CHAPTER 8 SECTION 3

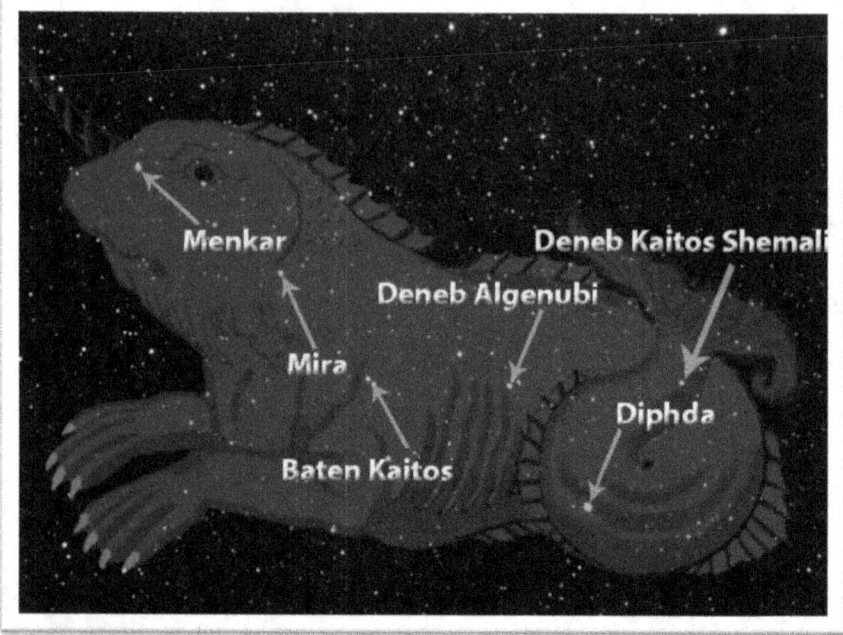

The second Decan in the house of Aries is Cetus. Its name in relation to the ancient zodiac of Dendera is *Knem*, meaning *subdued*. Cetus is none other than an image of God's arch enemy. He is also pictured in the heavens as the enemy of the Church and Israel; for chained to his neck are the bands of the two fish (Pisces). Furthermore, Cetus is seen trying to stop *Eridanus,* the river of blessing for the saints of God, and is, therefore, the enemy of Pisces and Andromeda.

PRINCIPAL STARS

The first star in Cetus worthy of discussing is *Mira*. Mira means *the rebel*. This star is said to appear and reappear, thus it has its name. This constellation is the perfect picture of the war between Lucifer and God. His rebellion is seen in manifold verses throughout the Bible. These angels were the first creation of God that we know of, and according to Job 38:7, they were present when God began his work here on the earth. Many of the angels were led into a rebellion by Lucifer. This *angelic rebellion* has been going on ever since (please see Jude 1:6; Matt. 25:41;

2nd Pet. 2:4; 1st Jn. 3:8; and Job 4:18). The good angels like to see that God is merciful, loving and kind.   See verses below.

- Luke 15:10, "Likewise, I say unto you, there is joy in the presence (face) of the angels of God over one sinner that repenteth."
- Ephesians 3:9-10, "And to make all *men* see what *is* the fellowship of the mystery, which from the beginning of the world hath been hid in God, who created all things by Jesus Christ: To the intent that now unto the principalities and powers in heavenly *places* might be known by the church the manifold wisdom of God."
- 1st Peter 1:12, "Unto whom it was revealed, that not unto themselves, but unto us they did minister the things, which are now reported unto you by them that have preached the gospel unto you with the Holy Ghost sent down from heaven; which things the angels desire to look into."
- Colossians 1:16, "For by him were all things created, that are in heaven, and that are in earth, visible and invisible, whether *they be* thrones, or dominions, or principalities, or powers: all things were created by him, and for him."
- Colossians 1:20, "And, having made peace through the blood of his cross, by him to reconcile all things unto himself; by him, *I say,* whether *they be* things in earth, or things in heaven."
- 1st Peter 3:22, "Who is gone into heaven, and is on the right hand of God; angels and authorities and powers being made subject unto him." God must be Lord universally (Phil. 2:10-11).

The next star in Cetus that is worthy of discussing is *Menkar*. Menkar means "the bound." Although the great enemy of God's kingdom is not bound in the bottomless pit for a thousand years yet, he is still bound in many respects. To the degree that Christians obey and follow Christ in righteousness, the less control and inroad the Devil has in their lives. The only inroad that Satan has in our lives is the one that we or God give him.

Sometimes, as in the case with Job, God allows Satan an inroad to our lives for testing, a higher purpose or to bring about something good; however, this is never done with evil, in the sense of sin (James 1:13). Job's Leviathan is pictured in the heavenly Cetus. God allowed Leviathan who is "king over all the children of pride" (Job 41:34), to tempt and afflict Job. As Job stood strong in his hour of trial, it is also our duty to not sway when under the fiery trial. Cetus' future is told in several passages. Consider Psalms 74:12-14, "For God *is* my King of old, working salvation in the midst of the earth. Thou didst divide the sea by thy strength: thou didst break the heads of the dragons in the waters. Thou didst break the heads of leviathan in pieces, *and* gavest him *to be* meat to the people inhabiting the wilderness."

This Breaker is seen in the next Decan of Aries, namely Perseus. One day, the great enemy of God and the human race will be destroyed. The Bible says in Revelation 20:10, "And the devil that deceived them was cast into the lake of fire and brimstone, where the beast and the false prophet *are,* and shall be tormented day and night forever and ever." Until then, the Church must be obedient, faithful and righteous for the Savior.

The third star in Cetus deserving of our attention is *Diphda*. Diphda means *the overthrown*. Satan is an overthrown angel, and his kingdom is defeated as well. This was accomplished at Calvary. The reason we have to look forward to these great events is because God's plan for the ages is still ongoing. So, in this powerful image in the night time sky, we see the Leviathan of the Bible and his future destruction. Isaiah 27:1, "In that day the LORD with his sore and great and strong sword shall punish leviathan the piercing serpent, even leviathan that crooked serpent; and he shall slay the dragon that *is* in the sea."

The stars of Cetus are still speaking to us today. They are telling us that Satan has no power over God; therefore, he is bound. Cetus is a reminder to the Christian not to "give place to the devil" (Eph. 4:27). As long as we stay under God's watch and care, Satan is as good as bound in our lives, but if we step out of bounds, we give place for the devil to work. In 1st Peter 5:8-9, we are admonished to "Be sober, be vigilant; because your adversary the devil, as a roaring lion, walketh about, seeking whom he may devour: Whom resist steadfast in the faith, knowing that the same afflictions are accomplished in your brethren that are in the world."

PERSEUS (THE BREAKER)
CHAPTER 8 SECTION 4

"The breaker is come up before them." (Micah 2:12-13)
The third decan in the house of Aries is Perseus. In Perseus, we see another image of Jesus Christ. Before we discuss the theme of Perseus, let's first look at the principal star names found within this decan.

PRINCIPAL STARS
- *Athik* meaning "who breaks."
- *Mirfak* meaning "who assists or helps."
- *Al Genib* meaning "who carries away."

IN THE HEAD CARRIED BY PERSEUS
- *Medusa* meaning "the trodden under foot."
- *Rosh Satan* meaning "the head of the enemy"
- *Al Oneh* meaning "the subdued, weakened."

- *Al Ghoul* meaning "the evil spirit."
- *Al Gol* is a changeable star in the head meaning "the coming and the going." Al Gol reminds us of Satan's ability to attack in stealth. At one time he can be a roaring lion (1st Pet. 5:8), another time as subtle as a serpent (Gen. 3:8), and yet at another time an angel of light (2nd Cor. 11:14).

In Micah 2:12-13, God informed Israel that he, their Messiah, would break through every obstacle in the way of their future restoration.

> I will surely assemble, O Jacob, all of thee; I will surely gather the remnant of Israel; I will put them together as the sheep of Bozrah, as the flock in the midst of their fold: they shall make great noise by reason of *the multitude of* men. The breaker is come up before them: they have broken up, and have passed through the gate, and are gone out by it: and their king shall pass before them, and the LORD on the head of them.

Likewise, before the great and destructive period of the Tribulation comes, God will rescue his bride from the destructive power of Satan and set free his Church (Andromeda).

THE MYTH

Perseus was the son of a divine Father. His birth was miraculous, in that a shower of gold descended upon Danae, which brought about his conception. Shortly after his birth, he was put under persecution. After overcoming his trial, he decided to bring a present to his father, namely the head of Medusa (a part of a three headed monster), who, along with his two other counterparts, defiled the temple of Zeus. While Perseus was on his way back from his noble deed, he saw Andromeda in chains at Joppa, ready to be devoured by Cetus. Perseus, then broke Cetus and freed Andromeda.

THE TRUTH BEHIND THE MYTH

Our Perseus, Jesus Christ, was born of a miraculous birth. "Therefore the Lord himself shall give you a sign; Behold, a virgin shall conceive, and bear a son, and shall call his name Immanuel" (Isaiah 7:14). Shortly after his birth, Christ was put under persecution by Herod's decree to kill the children two years old and under (Matt. 2:16). After defeating sin, death,

Hell and Satan on Calvary, Christ then "led captivity captive" (Ps. 68:18), freeing the Old Testament saints from Abraham's Bosom (Lk. 16:22). Psalms 68:18 is also referenced in Ephesians, "Wherefore he saith, When he ascended up on high, he led captivity captive, and gave gifts unto men. Now that he ascended, what is it but that he also descended first into the lower parts of the earth? He that descended is the same also that ascended up far above all heavens, that he might fill all things." (Ephesians 4:8-10).

The stars in Perseus are still speaking to us today. They are telling us that Christ will deliver the Old Testament and New Testament saints and establish his eternal covenant with Israel and her spiritual seed. After Satan has been bound in the bottomless pit for a thousand years, he will be loosed for his *final rebellion*, at which time our Perseus will have Satan cast into Hell along with his two other counterparts, namely, the beast and the false prophet. "And the devil that deceived them was cast into the lake of fire and brimstone, where the beast and the false prophet *are*, and shall be tormented day and night for ever and ever" (Revelation 20:10). Here, we see the culmination of Satan's plot being foiled. He subjugated the human race in the Garden of Eden ushering Genesis 3:16's proto-evangelism. Here, we can see the finality of Satan's rebellion.

Stir your heart dear saint of God
When on your turf, the devil's feet seem too trod.
Read the last chapters and see his end,
For all in Christ Jesus can shout, we win!

"I'm on the Winning Side"

Once I drifted out in sin,
had no hope nor joy within,
And my soul was burdened down with pride.
Then my Savior came along,
and He showed me I was wrong,
and now I know I'm on the winning side.

(Chorus)
Yes, I'm on the winning side;
Yes I'm on the winning side;
Out in sin no more will I abide.
I've enlisted in the fight for the cause of truth and right;
Praise the Lord – I'm on the winning side.

I will never have a fear,
For my Lord is ever near,
And in Him so often I confide;
He's the keeper of my soul
since I gave Him full control,
And He placed me on the winning side.

(Chorus)
Yes, I'm on the winning side;
Yes I'm on the winning side;
Out in sin no more will I abide.
I've enlisted in the fight for the cause of truth and right;
Praise the Lord – I'm on the winning side.

(Hal Reeves)

CHAPTER 9 SECTION 1

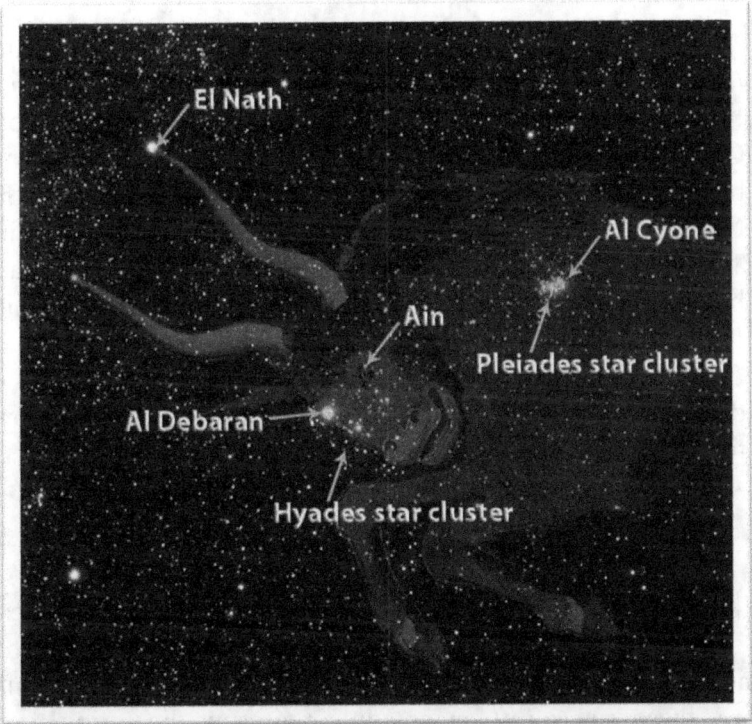

The Hebrew name of Taurus is שׁוּר, pronounced in English as *Shur*. This word means "the coming." In the Old Testament LXX, we see the Greek word *Tauros* in Deuteronomy 33:17. Taurus, like the other 11 main constellations, has three decans; *Orion, Eridanus*, and *Auriga*. Taurus presents an interesting chapter in the story of *Biblical Astrology*.

HISTORICAL ANALYSES OF THE WILD OX

The biblical unicorn is mentioned six times in the Old Testament and always refers to an extinct wild ox with two horns and not the one-horned mystical unicorn made popular today. Historical records by Julius Caesar describe this beast being hunted in the Hercynian forests in his time. Like behemoth and leviathan (Job 39:9-10), it also was aggressive toward humans. This unicorn was a large, two-horned ox that Caesar went on to say, "when a hunter succeeded in killing one (pitfalls being the chief means of capture), he made a public exhibition of the horns as the trophies of his success, which was the wonder and praise of all who beheld."[49]

PRINCIPAL STARS

- *Al Debaran* is located in the bull's eye and means *the captain* or *leader*. Here, we see the leader, captain, and father of the Kingdom of Darkness. This heavenly image does not depict Christ, but his counterfeit. Throughout the heavenly images, we have seen the repetitious story of Satan bruising Christ's heel and Christ bruising Satan's head. Likewise, we see this in the image of Taurus' horns piercing Auriga's heel, while Perseus is treading the head of Taurus. This is represented in the star *Athik*, meaning, *who breaks*.
- *El Nath,* as in Aries, means *the wounded*. This star is demonstrating the death blow to Auriga. It is delivered by one of the horns of Taurus striking the heel of Auriga.
- *Al Cyone*, located in the Pleiades cluster, means *the center*.

PLEIADES

God said to Job, "canst thou bind the sweet influences of Pleiades?" (Job 38:31). The words "sweet influences" are the Hebrew word מַעֲדַנּה pronounced in English as *mah-ad-an-naw,* which means influence. Therefore, these *sweet influences* could represent the luring power that Satan has in the world. This enticement sends men to the congregation of Hell by the millions. Only Jesus Christ is strong enough to break the power of Satan's sweet influences. Many modern Christians have a reversion philosophy which says, "now that you're saved, you can have Christ and live and adapt like the world." Although Christ is all powerful, he never stops others from going after the world.

HYADES

Hyades, a star cluster in Taurus, means *the Congregated*. The Bible mentions Hyades in eleven places in the Textus Receptus. In ten out of the eleven times it is used, it is translated as *hell* and in only one place it is translated grave. Therefore, Hyades is the place of the departed dead which are awaiting the Great White Throne Judgment. In every place, it refers to a holding place for those that will pay for their sin in the lake of

---

[49] Joseph Seiss, *The Gospel in the Stars,*

112

fire and brimstone. (See Matt. 11:23; 16:18; Lk. 10:15; 16:23; Acts 2:27, 31; 1st Cor. 15:55; and Rev. 1:18; 6:8; 20:13-14).

> And I saw a great white throne, and him that sat on it, from whose face the earth and the heaven fled away; and there was found no place for them. And I saw the dead, small and great, stand before God; and the books were opened: and another book was opened, which is *the book* of life: and the dead were judged out of those things which were written in the books, according to their works. And the sea gave up the dead which were in it; and death and hell (Hyades) delivered up the dead which were in them: and they were judged every man according to their works. And death and hell (Hyades) were cast into the lake of fire. This is the second death. And whosoever was not found written in the book of life was cast into the lake of fire. (Revelation 20:11-15)

Riding upon the shoulder of the wild bull is this congregation of the deceived. They are the deceived, in that they have allowed the Devil to convince them through sweet influences that they did not need Christ. They are deceived, in that they were allured by the world's charm and chose to reject Jesus Christ. They are deceived, in that they were convinced by the world's occults that downplayed Christ and magnified religion. As God has his congregation of *the redeemed* (the beehive cluster), so here, Satan has his congregation of *the condemned*. Upon the shoulder of this mighty powerful image of the great archenemy of God, he carries his congregated followers as a trophy of his power in the face of God. And it is us that are able to envision the finality of those bound in Hyades, the place of the dead, which from this time on, have "no hope and are without God" (Eph. 2:12).

113

Standing out in the middle of the night sky stands a constellation full of luminary orbs unlike no other, that when learned, the story is never forgotten. Although the myths of Greek Astrology have done their best to dirty up the waters of this once pure image of the invisible God, the message of Christ is still available for the student of God's Handiwork and Biblical Astrology. Orion is the first decan of Taurus, and is validated by God's Word in three places. It first appears in Job 9:9, "Which maketh Arcturus, Orion, and Pleiades, and the chambers of the south." Secondly, it appears in Job 38:31, "Canst thou bind the sweet influences of Pleiades, or loose the bands of Orion?" Thirdly, Orion is mentioned in Amos 5:8-9, "Seek him that maketh the seven stars and Orion..." There is little doubt that many of the ancient Bible writers were students in Biblical Astrology. Among them were Job and Amos. The Bible declared in Genesis 1:14 that the stars were to be "for signs." These signs were revealed through the study of the star's names. These names were given to the stars by God himself (Ps. 147:4; Is. 40:26) and have been preserved through the course of human history (Ecc. 3:14). The Bible said in Psalms 8:3, that God "ordained" the stars; this tells us that they have purpose. There is no doubt that stars were the declaration of God's love to the ancient world, as the Word of God is to us today. In the story of Orion, the message retrieved from the stars articulates and complements the gospel story found in the Scriptures. Orion's name is interpreted as *the Bearer of Light*. This is significant in the story of God, in that Christ said on several occasions that he was "the light of the world" (John 8:12; 9:5; 12:46). It is said in Revelation 21:23, concerning the eternal state with God, that there will be no need of the sun or moon "for the glory of God did lighten it, and the Lamb is the light thereof." Since the story of the stars is revealed through the interpretations of their names, let's look at some of the names of the stars in Orion and see if they enlighten the Bible's story.

BETELGUESE "THE COMING BRANCH"

The first star we will observe is *Betelgeuse*. Betelgeuse is a star of the first magnitude on Orion's right shoulder and means "the coming Branch." The Old Testament writers, in more places than one (Zech. 3:8; 6:12; Jer. 23:5-6; 33:15), penned under inspiration that there was to come a Branch (Messiah) from the seed of Jesse (Is. 11:1-5). This prophecy was fulfilled in Christ and validated in the New Testament (Matt. 1:6-16). In Revelation 22:16 the Bible says, "I Jesus have sent mine angel to testify unto you these things in the churches. I am the root and the offspring of David, and the bright and morning star." The Greek word for "root" is ῥίζα, which is the same underlying word in the Septuagint (LXX) for *stem* and *roots* found in the Isaiah 11:1's *Coming Branch Prophecy*. This signifies that Christ is the fulfillment of these Old Testament Branch prophecies.

ORION'S BELT

Signifying *the Coming Branch*, Isaiah 11:1 goes on to tell us in verse five that the Branch would have "righteousness for the girdle of his loins and faithfulness for the girdle of his reigns" which symbolizes beautifully *Orion's Belt*. Orion's Belt has three beautiful shining brilliants, representing Christ's girdle (Is. 11:1-5). This belt has been a wonder for many stargazers. These stars typify the implacable suffering and sacrifice that Jesus was to go through in order to tip the balance scales of God's justice for humanity's sin. These stars also resemble the great love our

King has for his kingdom. No wonder so many have marveled at their beauty for thousands of years.

PRINCIPAL STARS

Rigel is a star of the first magnitude flaming in Orion's lifted left foot. Rigel means "the foot that crusheth." If you look under the left foot of Orion, you will see **Lepus**, *the hare* (the first decan of Gemini). In this story Lepus is a picture of Satan and his principal stars mean "the mad, the caught, and the deceiver." Once again, this picture is a very familiar message among the heavenly bodies, which typifies the greatest and oldest prophecy in the Scriptures located in Genesis 3:15. "And I will put enmity between thee and the woman, and between thy seed and her seed; it shall bruise thy head, and thou shalt bruise his heel." In this prophecy, we see that through the seed of the woman would come the One that would bruise the Serpent's head in order to redeem fallen man. In order to accomplish this objective there would be a heavy price to pay. This payment is symbolized in some of Orion's stars, such as, *Al Nitak*, which means "the wounded," and *Saiph*, which means "the bruised." The Bible says in Isaiah 53:5 that Christ, "was <u>wounded</u> for our transgressions and <u>bruised</u> for our iniquities." This would all become fulfilled in the Crucifixion of Christ on Calvary. This was the bruising of the heel mentioned in Genesis 3:15, showing the struggle that Christ had to endure in order to triumph over the Devil, sin, and death.

BELLATRIX

Located on Orion's left breast is a bright star called *Bellatrix*, which means "swiftly coming." Swiftly Christ came in to this world, in his First Advent.[50] By the time the forces of evil could catch up with him, he was at Calvary laying down his life for the sin of humanity. In Orion's right hand he holds a wooden club, and in his left he holds the carcass of a lion. It was at the Cross where the Devil, who "as a roaring lion seeking whom he may devour" (1 Peter 5:8) met his match; and although he delivered a death blow to the heel of Christ that day, our Glorious One took his head! This is why there is no struggle in the end to cast him in the lake that burns with fire and brimstone, because the struggle was already accomplished at Calvary.

---

[50] The First Advent is a doctrine referring to Christ's first coming found in the gospels.

THE MYTHS

The Astrology myths, though full of contradictions, say that Orion had power to walk on the sea without getting his feet wet. He surpassed all others in handsomeness, strength, character, and stature. He was the greatest hunter, capable of conquering every animal on earth. Because of this claim, Lepus came up out of the earth and gave him a mortal wound in his foot, but at Diana's request he was raised to immortality, and placed in the heavens over against *the Hare* (Lepus). It is said that because he loved Merope, her father put out his eyes while he was asleep on the sea shore, but that by raising himself on the back of a blacksmith and turning his face to the rising sun, he recovered his sight and went forth with great haste, rage and energy to avenge the perfidious cruelty of his foes. Later he is said to have loved the Pleiadic (gentile) maiden and that out of affection for her he performed the great work of clearing the country of all harmful and wild beasts, bringing the spoils of his successes as presents to his beloved.

THE MYTHS IN LIGHT OF CHRIST

God, having his first love for the Jews, was horrified by their fornications (perfidious ways) and set his eyes on a redeemed Gentile Bride. Like Orion, Christ was born a peculiar gift of deity (Matt. 2:11). He was indeed the greatest of all men. He was in the world without being wetted or soiled by its waters (Heb. 4:15). Because of his love for the Church, he sunk into a deep sleep upon Calvary (Jn. 3:16), in which, the light of his eyes was extinguished at death, but revived at his resurrection (Jn. 11:25). He came into the world to destroy all the mighty powers of evil, and on this account he was stung in the heel by Satan. He now has supreme power in Heaven and earth (Matt. 28:18-20) and is stationed in Immortal glory at the right hand of the Father.

> All hail the power of Jesus' name, let angels prostrate fall,
> Bring forth the royal diadem, and crown him Lord of all!
> (Edward Perronet)

The prophet Amos said it well in Amos 5:6-8, when he said, "Seek the Lord....that maketh the seven stars and Orion..." Dear reader are you seeking him?

ORION

Daniel G. McCrillis
Daniel G. McCrillis

Score

Piano

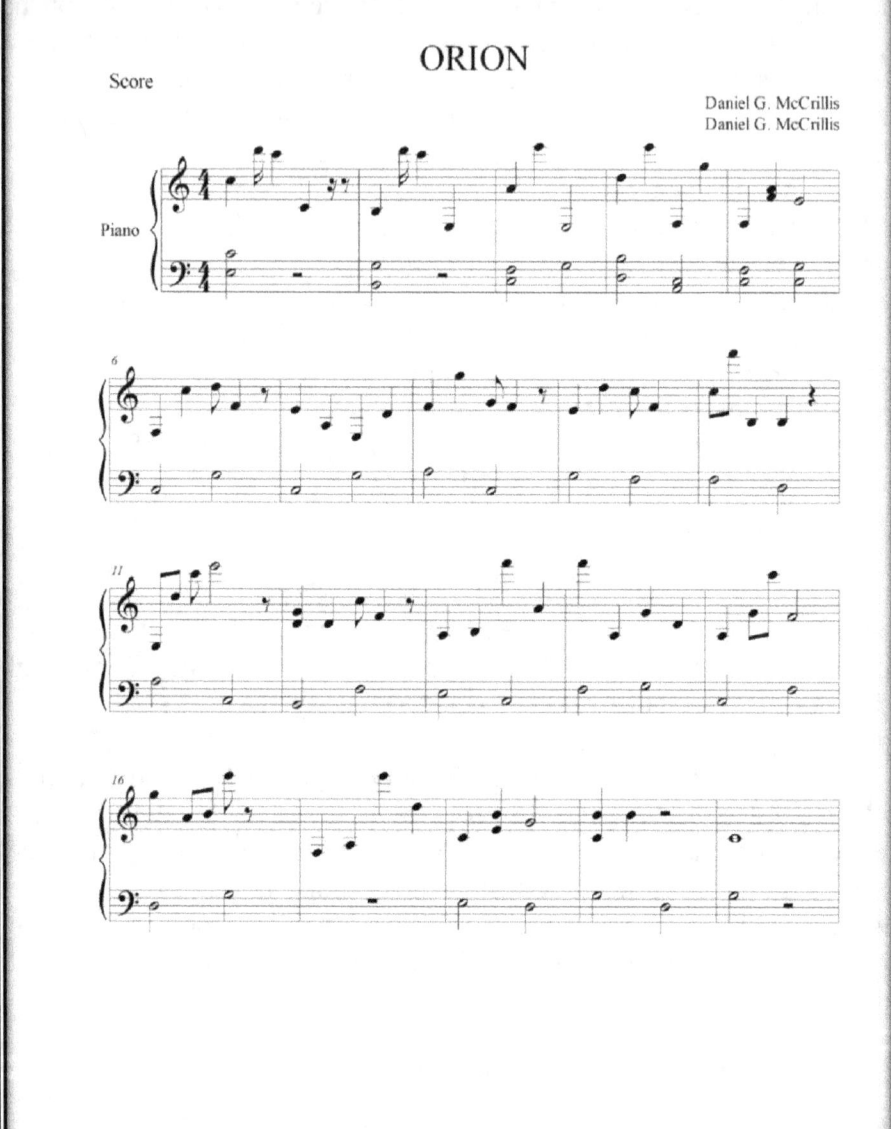

ERIDANUS (THE RIVER OF LIFE)
CHAPTER 9 SECTION 3

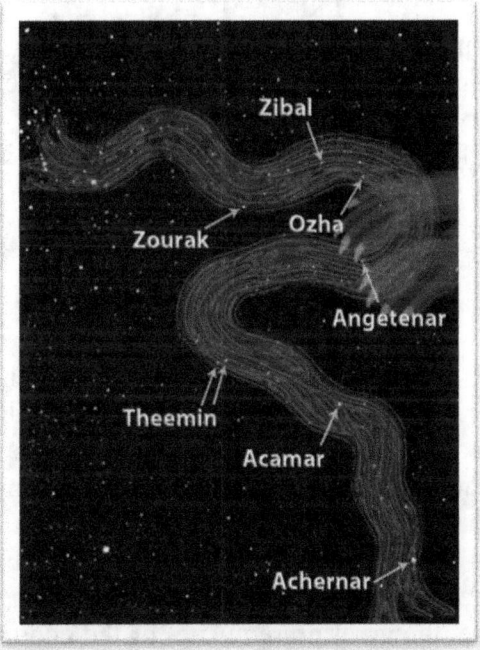

Eridanus is the second decan of Taurus. Many of the older works of *Biblical Astrology* have patterned *Eridanus* after the Greek perversion by making it a river of fire or indignation. As far as the names of the stars are concerned, none of them interpret fire in any way. Therefore, this river is one of life and blessing, not damnation. This river begins at the left foot of Orion as he completes the Genesis 3:15 gospel; getting the head of the enemy. This star is called *Rigel* and means the foot which crushes, which pictures Calvary's power and victory. The heavenly picture here signifies that Satan's hold on this world and our lives was broken at Calvary. It was because of Calvary that the life of blessing begins. Likewise, it is here that this most wondrous constellation begins. This river is of life and blessing for all those who have come under the redemptive plan of God.

Revelation 22:1-2, And he shewed me a pure river of water of life, clear as crystal, proceeding out of the throne of God and of the Lamb. In the midst of the street of it, and on either side of the river, *was there* the tree of life, which bare twelve *manner of* fruits, *and*

119

yielded her fruit every month: and the leaves of the tree *were* for the healing of the nations.

- *Achernar* means "the part of a river."
- *Phaet* means "mouth of the river." Where the river flows into a lake, reservoir, basin, ocean or sea. Death is the end of the beginning for the believer, letting us know it is the entrance into eternal blessing, not the end of a temporal opportunity.
- *Theemin* means "the water."
- *Ozha* means "the going forth."
- *Zourak* means "flowing."

It is noteworthy to mention that Cetus the sea monster is vainly trying to stop the flow of Eridanus. Likewise, he is out to stop the blessings of God in the lives of God's children. The stars in Eridanus are speaking to us today; they are teaching us that **death** has no more power, and what was lost in the Garden of Eden is restored in Jesus Christ. God assured this truth in the giving of his Holy Spirit in the life of the believer. **Satan** has no more power over the believer, only that which the believer permits him to have. Therefore, the great prerequisite for the continued blessing upon the life of the Christian is the believer's trust and obedience.

Psalm 1 Blessed *is* the man that walketh not in the counsel of the ungodly, nor standeth in the way of sinners, nor sitteth in the seat of the scornful. But his delight *is* in the law of the LORD; and in his law doth he meditate day and night. And he shall be like a tree planted by the rivers of water, that bringeth forth his fruit in his season; his leaf also shall not wither; and whatsoever he doeth shall prosper. The ungodly *are* not so: but *are* like the chaff which the wind driveth away. Therefore the ungodly shall not stand in the judgment, nor sinners in the congregation of the righteous. For the LORD knoweth the way of the righteous: but the way of the ungodly shall perish.

Psalms 84:11-12, "For the LORD God *is* a sun and shield: the LORD will give grace and glory: no good *thing* will he withhold from them that walk uprightly. O LORD of hosts, blessed *is* the man that trusteth in thee."

Revelation 22:17, "And the Spirit and the bride say, Come. And let him that heareth say, Come. And let him that is athirst come. And whosoever will, let him take the water of life freely."

"Trust and Obey"

When we walk with the Lord in the light of His Word,
What a glory He sheds on our way!
While we do His good will He abides with us still,
And with all who will trust and obey.

(Refrain) Trust and obey—for there's no other way
To be happy in Jesus, but to trust and obey.

Not a shadow can rise, not a cloud in the skies,
But His smile quickly drives it away;
Not a doubt or a fear, not a sigh or a tear,
Can abide while we trust and obey. (Refrain)

Not a burden we bear, not a sorrow we share,
But our toil He doth richly repay;
Not a grief or a loss, not a frown or a cross,
But is blessed if we trust and obey. (Refrain)

But we never can prove the delights of His love
Until all on the altar we lay;
For the favor He shows and the joy He bestows,
Are for them who will trust and obey. (Refrain)

Then in fellowship sweet we will sit at His feet.
Or we'll walk by His side in the way.
What He says we will do, where He sends we will go;
Never fear, only trust and obey. (Refrain)

By John H. Sammis

AURIGA (THE SHEPHERD)
CHAPTER 9 SECTION 4

Auriga is represented as an image of a man who is seated on top of the Milky Way Galaxy. He is called the Shepherd, and his right hand has let go of the bands that go to Pisces (the Fishes). Seated in his lap is a baby lamb or she goat. This image is a most magnificent picture of Christ as Shepherd. In Isaiah 40:11, the Bible says, "He shall feed his flock like a shepherd: he shall gather the lambs with his arm, and carry *them* in his bosom, *and* shall gently lead those that are with young."

PRINCIPAL STARS

- *El Nath* means "wounded."
- *Maaz* means "flock of goats."
- *Menkalinon* means "band or chain."
- *Alioth* means "she goat."

In this image, Auriga's heel is being bruised by the horn of Taurus (Gen. 3:15). In the most ancient zodiac of Dendera (2,500 B.C.), Auriga holds a scepter, the top showing the head of a lamb and the bottom showing a cross. From the earliest of time until now, the signs that are derived from the stars' names have told one in the same message.

The stars of Auriga are still speaking to us today. They are reminding us of: God's universal sovereignty over all creation, the believer's safety that he has in Christ, and of the great love that the Shepherd has for the sheep.

O, the love that drew salvation's plan!
O, the grace that brought it down to man!
O, the mighty gulf that God did span, at Calvary!
(William Reed Newell 1868-1956)

GEMINI "THE TWINS"
CHAPTER 10 SECTION 1

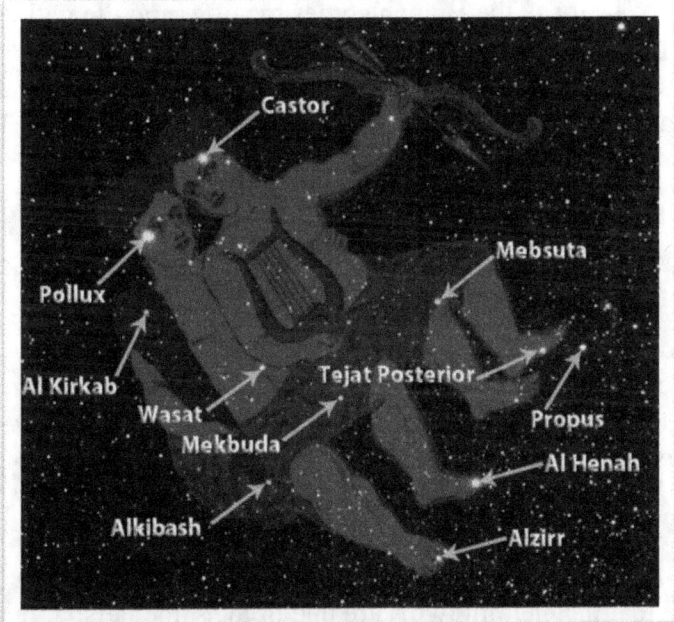

Acts 28:11 "And after three months we departed in a ship of Alexandria, which had wintered in the isle, whose sign was Castor and Pollux." The Hebrew name for Gemini is pronounced in English as Thaumim, and means *the united*. It is likewise interpreted in the Syriac and Coptic. Gemini, like the other 11 main constellations, has three decans; *Lepus*, *Canis Major*, and *Canis Minor*, and brings out some amazing truths to the gospel story found within the stars.

THE PRINCIPAL STARS

- *Propus*, meaning "the branch."
- *Mebsuta*, meaning "the treading under foot."
- *Wasat*, (Arab) meaning "set or appointed."
- *Al Giauza*, meaning "the palm branch or stem."
- *Al Henah*, (located in the foot of Pollux) "meaning hurt or afflicted."
- *Al Dira*, meaning "the seed or branch."
- *Castor*, (Apollo bearing an arrow) meaning "ruler or judge."
- *Pollux*, (Hercules) meaning "coming to suffer."

125

The two brightest stars in Gemini are Castor and Pollux. Castor means *ruler*, and Pollux means *coming to suffer*. With these stars and their locations in the heavens, we see the dual nature of Christ. We see the Incarnation of Deity as the Branch comes to suffer and die with the star *Al Henah*. This is significant in that it also is in the foot of Pollux, not Castor. "And I will put enmity between thee and the woman, and between thy seed and her seed; it shall bruise thy head, and thou shalt bruise his heel." (Gen. 3:15)

THE MYTHOLOGY

In Greek mythology, the twins were Apollo and Hercules, the divine sons of Zeus. In their most notable exploit, they cleared the seas of all the pirates; and it was because of the myth that we see a ship in Acts 28:11 which bore their Latin names of Castor and Pollux, thus making a superstitious good luck charm for safe voyages, out of a powerful truth tucked away in the stars.

THE TRUTH IS STILL VISIBLE

The stars in Gemini are still speaking to us today. Through their names, one can see that the Gemini twins are one in the same Person, representing Christ's deity and humanity. It was through this dual nature that Christ was able to come in humility and be exalted to glory. The Bible informs us in Jeremiah 23:5-6 that "Behold, the days come, saith the LORD, that I will raise unto David a righteous Branch, and a King shall reign and prosper, and shall execute judgment and justice in the earth. In his days Judah shall be saved, and Israel shall dwell safely: and this is his name whereby he shall be called, THE LORD OUR RIGHTEOUSNESS." From Gemini, we can see that the heavenly sign is a symbol of the dual nature of the Coming Messiah. His plan was executed at Calvary, where he provided a blood-bought redemption for every sinner. The sea is mentioned in the Bible as a symbol of the world. Like Castor and Pollux, our Lord, through his dual nature, came into the world to rid it from all its sinful vices. He was in all points tempted as humanity, "yet without sin" (Heb. 4:15). "He was in the world, and the world was made by him, and the world knew him not. He came unto his own, and his own received him not: But as many as received him, to them gave he power to become the sons of God, *even* to them that believe on his name: Which were born, not of blood, nor of the will of the flesh, nor of the will of man, but of God" (Jn. 1:10-13).

"At Calvary"

Years I spent in vanity and pride,
Caring not my Lord was crucified,
Knowing not it was for me He died on Calvary.

(Refrain) Mercy there was great, and grace was free;
Pardon there was multiplied to me;
There my burdened soul found liberty at Calvary.

By God's Word at last my sin I learned;
Then I trembled at the law I'd spurned,
Till my guilty soul imploring turned to Calvary. (Refrain)

Now I've given to Jesus everything,
Now I gladly own Him as my King,
Now my raptured soul can only sing of Calvary! (Refrain)

O, the love that drew salvation's plan!
O, the grace that brought it down to man!
O, the mighty gulf that God did span at Calvary! (Refrain)

(William Reed Newell 1868-1956)

LEPUS "THE HARE"
CHAPTER 10 SECTION 2

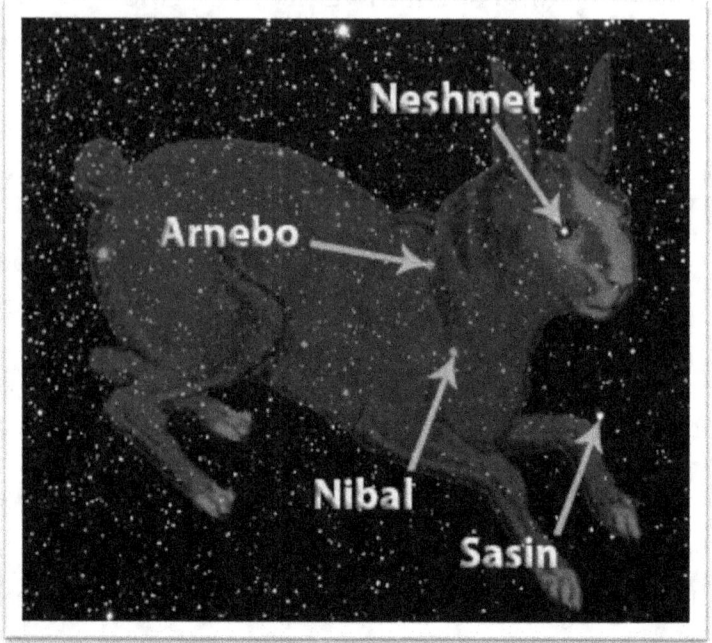

"I have trodden the winepress alone; and of the people there was none with me: for I will tread them in mine anger, and trample them in my fury; and their blood shall be sprinkled upon my garments, and I will stain all my raiment. For the day of vengeance is in mine heart, and the year of my redeemed is come." (Isaiah 63:3-4)

The modern name for Lepus is from the Latin and Roman version of the stars. In the zodiac of Denderah, the image is called *Bashti-beki* and pictured as an unclean bird standing on a serpent. In spite of the differences between Latin and Egyptian images, the interpretation of the stars tells one in the same story.

THE PRINCIPAL STARS
- *Arnebo*, (Heb) meaning "the enemy of him that cometh."
- *Nibal*, meaning "the mad."
- *Rakis*, (Arab) meaning "the bound."
- *Sugia*, meaning "the deceiver."

Under the feet of Orion, we see in the constellation of Lepus another fulfillment of Genesis 3:15. Genesis 3:15 "And I will put enmity between thee and the woman, and between thy seed and her seed; it shall bruise thy head, and thou shalt bruise his heel." Is it not odd that the signs in the heavens depict a story as repetitious as this? Evidently, the importance of Genesis 3:15 was so crucial to God that he, by divine providence and sovereign foreknowledge, placed it in the heavens numerous times. How one can spend any amount of time studying the stars and not see that Christ was the Lamb slain from the foundation of the world is beyond comprehension.

CANIS MINOR "THE SECOND DOG"
CHAPTER 10 SECTION 3

In Canis Major, we will see the deity of Christ, namely his majesty, but in Canis Minor, we see the humanity of Christ. This is the Christ that died on the Cross to become the Redeemer of the fallen race. This demonstrates the completeness and perfection of God, in that it took deity wrapped in humanity to accomplish God's highest objective. In Canis Major, we see the exalted Christ; but in Canis Minor, we see Christ carrying the load of Calvary. Let's look at these stars now.

THE PRINCIPAL STARS

- *Procyon*, (Heb) meaning "the redeemer."
- *Al Shira Al Shemeliya*, meaning "the chief of the left hand."
- *Al Mirzam*, (Arab) meaning "the prince or ruler."
- *Al Gomeyra*, meaning "who completes or perfects." What a wonderful attribute of God. Deuteronomy 32:4 says that "*He is* the Rock, his work *is* perfect: for all his ways *are* judgment: a God of truth and without iniquity, just and right *is* he." Psalms 18:30 says, "*As for* God, his way *is* perfect: the word of the LORD is tried: he *is* a buckler to all those that trust in him."
- *Al Gomeisa*, meaning "loaded and bearing for others." "Looking unto Jesus the author and finisher of *our* faith; who for the joy that was set before him endured the cross, despising the shame, and is set down at the right hand of the throne of God" (Heb 12:2).

THE CHIEF OF RIGHT AND LEFT HAND

In Canis Major, we saw the star *Al Shira Al Jemeniya*, meaning "the chief of the right hand" and in Canis Minor we have the star *Al Shira Al Shemeliya*, meaning "the chief of the left hand." It is important for the Christian to keep a proper balance in the right hand/left hand understanding of Christ. In this most unusual construction of the nighttime sky, we see the importance of two parts of Christ. Gemini has put this message across very well in Castor and Pollux, and here in its second decan, we see another picture of the dual nature of Christ. We as Christians should bear on our right hands that we have the Almighty King of kings and Lord of lords on our side (Canis Major), yet simultaneously, we should understand on our left hand that God has feelings, emotions, and humanity as we do (Canis Minor). Therefore, no matter how great God is, we should always understand that he is a Person. He wants to spend time with us. He wants to fellowship with us. He wants us on our right hand to get into the prayer closets and studies to receive the power of God, but on our left hand we should leave the prayer closet, taking our Savior with us, and live down the power of God through obedience. This is the God that we are to know. In Psalm 121:5, David told Israel that "the LORD is thy keeper: the LORD is thy shade upon thy right hand," signifying how close God is to them. We as Christians have this same inter-dispensational promise that God "will never leave thee nor forsake thee" (Heb. 13:5). One who does not understand the dual nature of Christ is a candidate for problems.

Canis Major
"The Dog" or Sirius "The Prince"
Chapter 10 Section 4

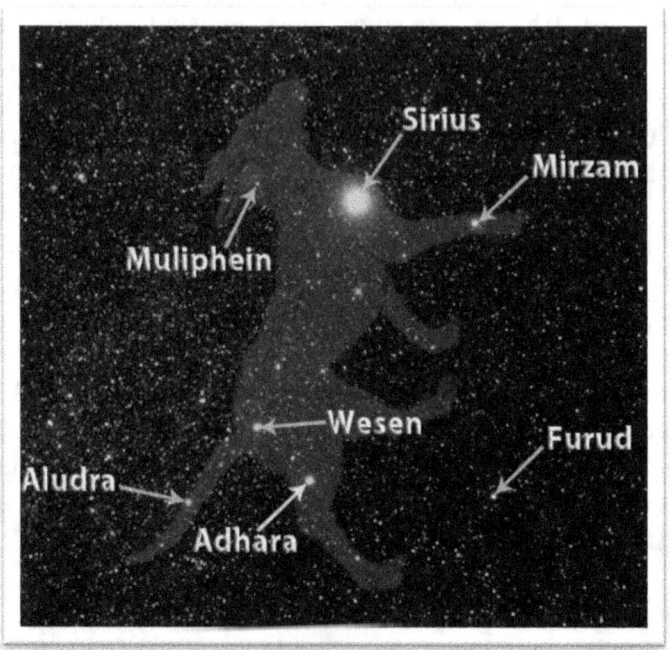

Isaiah 9:6 "For unto us a child is born, unto us a son is given: and the government shall be upon his shoulder: and his name shall be called Wonderful, Counsellor, The mighty God, The everlasting Father, The Prince of Peace."

The second decan in Gemini is known to the world as Canis Major, but to the ancient world, he is known by the brightest star in the heavens, namely Sirius. Sirius means the Prince, and we are reminded in Isaiah 9:6 that Christ is the "Prince of Peace." It is only evident that all of the stars in this constellation depict the awe inspiring majesty of Christ who sits at the right hand of God the Father (*Al Shira Al Jemeniya*), who came and is coming again (*Aschere*), who is the King of kings (*Mizram*) and the bright and morning Star (*Wesen*)!

THE PRINCIPAL STARS

- *Sirius*, meaning "the prince."
- *Aschere*, meaning "who shall come."
- *Al Shira Al Jemeniya*, (Arab) meaning "the chief of the right hand."
- *Al Habor*, meaning "the mighty."
- *Adhara*, meaning "the glorious."
- *Mirzam*, meaning "the prince or ruler."
- *Wesen*, meaning "the bright and shining."

All Hail the Power of Jesus Name

All hail the power of Jesus' Name! Let angels prostrate fall;
Bring forth the royal diadem, and crown Him Lord of all.
Bring forth the royal diadem, and crown Him Lord of all.

Let highborn seraphs tune the lyre, and as they tune it, fall
Before His face Who tunes their choir, and crown Him Lord of all.
Before His face Who tunes their choir, and crown Him Lord of all.

Crown Him, ye morning stars of light, who fixed this floating ball;
Now hail the strength of Israel's might, and crown Him Lord of all.
Now hail the strength of Israel's might, and crown Him Lord of all.

Crown Him, ye martyrs of your God, who from His altar call;
Extol the Stem of Jesse's Rod, and crown Him Lord of all.
Extol the Stem of Jesse's Rod, and crown Him Lord of all.

Ye seed of Israel's chosen race, ye ransomed from the fall,
Hail Him Who saves you by His grace, and crown Him Lord of all.
Hail Him Who saves you by His grace, and crown Him Lord of all.

Hail Him, ye heirs of David's line, whom David Lord did call,
The God incarnate, Man divine, and crown Him Lord of all,
The God incarnate, Man divine, and crown Him Lord of all.

Sinners, whose love can ne'er forget the wormwood and the gall,
Go spread your trophies at His feet, and crown Him Lord of all.
Go spread your trophies at His feet, and crown Him Lord of all.

Let every tribe and every tongue before Him prostrate fall
And shout in universal song the crownèd Lord of all.
And shout in universal song the crownèd Lord of all.

O that, with yonder sacred throng, we at His feet may fall,
Join in the everlasting song, and crown Him Lord of all,
Join in the everlasting song, and crown Him Lord of all!

Edward Perronet (1779) John Rippon (1787)

CANCER "THE SCARABAEUS"
CHAPTER 11 SECTION 1

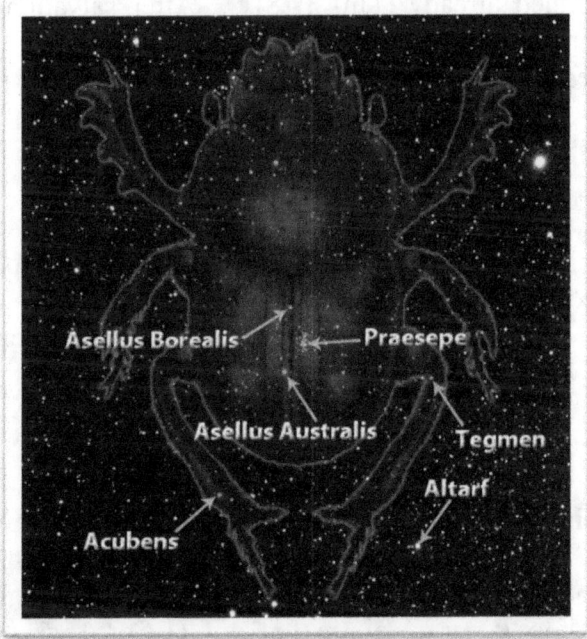

"Wherefore seeing we also are compassed about with so great a cloud of witnesses, let us lay aside every weight, and the sin which doth so easily beset us, and let us run with patience the race that is set before us." (Hebrews 12:1)

Cancer, like the other 11 main constellations, has three decans; *Ursa Major*, *Ursa Minor*, and *Argo*. Cancer brings forth a wonderful addition to the already amazing gospel story found in *Biblical Astrology*. In the Ancient Zodiacs of Denderah and Esneh, Cancer is not pictured as a crab, but as a Scarabaeus, the sacred Egyptian beetle. The Ancient Egyptians placed the *scarabaeus sacer* among their most sacred species in the world. This beetle comes into the world in the lowest of ways, in that; it is born in a ball of dung. After mating, the male and female beetle will prepare a brooding ball; in which, the female will lay her eggs. The Egyptians likened this process to Kephri, the early morning manifestation of the sun god Ra; of which, it was believed that Kephri's job was to roll the sun across the sky. Likewise, the *Scarabaeus sacer* is known to roll balls of dung across the ground to feed its offspring in an underground chamber.

The male can roll a ball up to 1,000 times its weight! Furthermore, the *Scarabaeus sacer* goes through the metamorphosis process. All of this beautifully depicts the Christian's life. He was born in the dung, as a sinner "condemned already" (Jn. 3:18).[51] After salvation he is a new creature in Christ Jesus. This new life is his assurance of an eternal life with God in heaven.

> Behold, I shew you a mystery; We shall not all sleep, but we shall all be changed, In a moment, in the twinkling of an eye, at the last trump: for the trumpet shall sound, and the dead shall be raised incorruptible, and we shall be changed. For this corruptible must put on incorruption, and this mortal *must* put on immortality. So when this corruptible shall have put on incorruption, and this mortal shall have put on immortality, then shall be brought to pass the saying that is written, Death is swallowed up in victory. O death, where *is* thy sting? O grave, where *is* thy victory? The sting of death *is* sin; and the strength of sin *is* the law. But thanks *be* to God, which giveth us the victory through our Lord Jesus Christ. Therefore, my beloved brethren, be ye stedfast, unmoveable, always abounding in the work of the Lord, forasmuch as ye know that your labour is not in vain in the Lord. (1 Cor. 15:51-58)

THE PRINCIPAL STARS

- *Praesepe,* (Hebrew) meaning "a multitude." In Latin it means "a manger." *Praesepe* is one of the nearest star clusters to the Solar System. *Praesepe* contains a larger than average star population than most other gravitationally bound clusters. Its nickname among astronomers and backyard observers is "the Beehive Cluster." Altogether the Beehive Cluster contains over 1,000 gravitationally bound stars. What is unique about this cluster is that its name, meaning a multitude, seems to indicate the same image that the rest of the constellation does, namely, the saints of God.
- *Ma'alaph,* meaning "assembled thousands."
- *Al Himarein,* (Arabic) meaning "the kids or lambs." This resembles what the cluster consists of; namely, the church and Israel.

---

[51] The Depravity of Man is a doctrine referring to the vitiated state of man inherited from the Fall and Satan's Subjugation of the human race.

These stars in Cancer tell a most wonderful story. As we saw the Hyades cluster in Taurus being the abode of the condemned, so here, we see the abode of the righteous. In Genesis 22:17, God informed Abraham that he would multiply his seed as the stars of heaven. "That in blessing I will bless thee, and in multiplying I will multiply thy seed as the stars of the heaven, and as the sand which *is* upon the sea shore; and thy seed shall possess the gate of his enemies" (Gen. 22:17). This seed went beyond Abraham's natural pro-generation, but included the seed of faith. "Now to Abraham and his seed were the promises made. He saith not, And to seeds, as of many; but as of one, And to thy seed, which is Christ" (Gal. 3:16). Paul goes on to explain, "And if ye *be* Christ's, then are ye Abraham's seed, and heirs according to the promise" (Gal. 3:29).

Here in Cancer, we see a most spectacular Nebula, just visible to the naked eye that reminds us of our peculiarity. It is reminding us that we are to look "for that blessed hope, and the glorious appearing of the great God and our Saviour Jesus Christ; Who gave himself for us, that he might redeem us from all iniquity, and purify unto himself a peculiar people, zealous of good works" (Titus 2:13-14). The stars in Cancer are speaking to us today. They are telling us of the peculiarity that we have with Christ.

The Beehive Cluster

Ancient astronomers have been enamored with the beauty of the Beehive Cluster. Ptolemy described it as, "a nebulous mass in the breast of Cancer."[52] Significantly, some astronomers have noticed its comparability to Hyades; indicating a correlation in origin, motion, star signatures and distance.[53] From earth and in spiritual terms Hyades would represent the congregation of the condemned and *Praesepe* the congregation of the redeemed.

[52] Messier 44: Observations and Descriptions (about 130 A.D.; *Almagest*).
[53] Klein—Wassink, WJ *The Proper Motion and the Distance of the Praesepe Cluster,* Publications of the Kapteyn Astronomical Laboratory Groningen, (1927).

As for the origin both contain white dwarfs and red giants. Spectrometry indicates that the spectral signatures are classes A, F, G, K, and M, which indicate the later stages of stellar creation.[54] Once Cancer's most ancient image is recognized as a scarab instead of a crab, the idea of its representation being evil vanishes and a marvel settles upon the mind with the Beehive Cluster. The wonder that forces its way to our attention is, "Is the Beehive Cluster where the souls of the redeemed are awaiting the redemption of their glorified bodies?"[55]

---

[54] Monthly Notices of the Royal Astronomical Society, New Praesepe White Dwarfs and the initial mass/final mass relation, Dobbe PD., Napiwotzki R; Burleigh MR, 2006.
[55] Rom. 8:23b …even we ourselves groan within ourselves, waiting for the adoption, *to wit*, the redemption of our body.

"Faith of our Fathers"

Faith of our fathers, living still,
In spite of dungeon, fire, and sword;
O how our hearts beat high with joy
Whene'er we hear that glorious word!

Refrain: Faith of our fathers, holy faith!
We will be true to thee till death.

Faith of our fathers, we will strive
To win all nations unto thee;
And through the truth that comes from God,
We all shall then be truly free.
(Refrain)

Faith of our fathers, we will love
Both friend and foe in all our strife;
And preach thee, too, as love knows how
By kindly words and virtuous life.
(Refrain)

By: Frederick W. Faber, 1814-1863

Ursa Minor
"The Sand Seed of Israel"
Chapter 11 Section 2

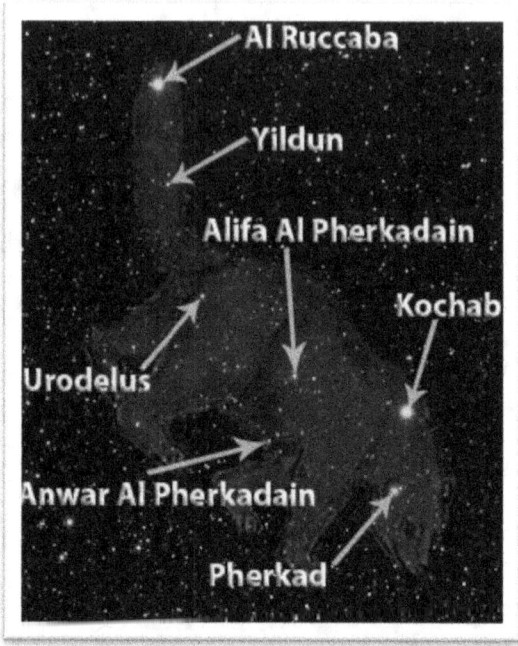

Job 38:32 "Canst thou bring forth Mazzaroth in his season? or canst thou guide Arcturus with **his sons**?"

Ursa Minor is known to most as "the Little Dipper." It is the Little and Big Dipper that are primevally known as the sons of Arcturus, better known as Bootes, the Oriental Shepherd. These sons are sheep folds or flocks, representing Israel and the Church. Although it is represented as a little bear in Greek and Roman astrology, none of the ancient zodiacs depict such an image. What is depicted in the Little and Great Bears are the sheep folds of God, namely the sand and star seed of Israel—Israel and the Church. After visiting Cancer and seeing the myriads of multitudes of the people of God, here we see another image of the same picture. In Genesis 22:17's blessing and covenant, we see the added statement, "and thy seed shall possess the gate of his enemies" which is often overlooked by covenant theologians. Not only is the covenant "the gate" for the seed of Abraham's enemies (whether by blessing or curse), but it is also in the

140

Little Dipper that we see the Pole-star, of which all other stars seem to turn; thus showing that this seed does most wonderfully possess the gate of his enemies.

THE PRINCIPAL STARS

- *Kochab,* (Hebrew) meaning "waiting him who cometh."
- *Al Pherkadain,* (Arabic) meaning "the calves young."
- *Al Gedi,* (Arabic) meaning "the kid."
- *Al Ruccaba,* (Arabic) meaning "the pole-star, the turned, or the ridden on."
- *Al Kaid,* (Arabic) meaning "the assembled" as in Ursa Major.

Since Abraham is "the Father of us all," we must first start with his seed and interpret Ursa Minor as the seed of Abraham prior to Calvary. This includes Jews and gentiles in the Old Testament, of which are: the mixed multitude (Ex. 12:38), Rahab (Heb. 11:31), the widow of Zarephath (I Kings 17:24), etc. In Ursa Minor, we have the eternal promise for the Sand of Israel. Romans 4:16 "Therefore it is of faith, that it might be by grace; to the end the promise might be sure to all the seed; not to that only which is of the law, but to that also which is of the faith of Abraham; who is the father of us all." "Now to Abraham and his seed were the promises made. He saith not, And to seeds, as of many; but as of one, And to thy seed, which is Christ" (Gal. 3:16). Paul goes on to explain, "And if ye *be* Christ's, then are ye Abraham's seed, and heirs according to the promise" (Gal. 3:29).

URSA MAJOR
"THE STAR SEED OF ISRAEL"
CHAPTER 11 SECTION 3

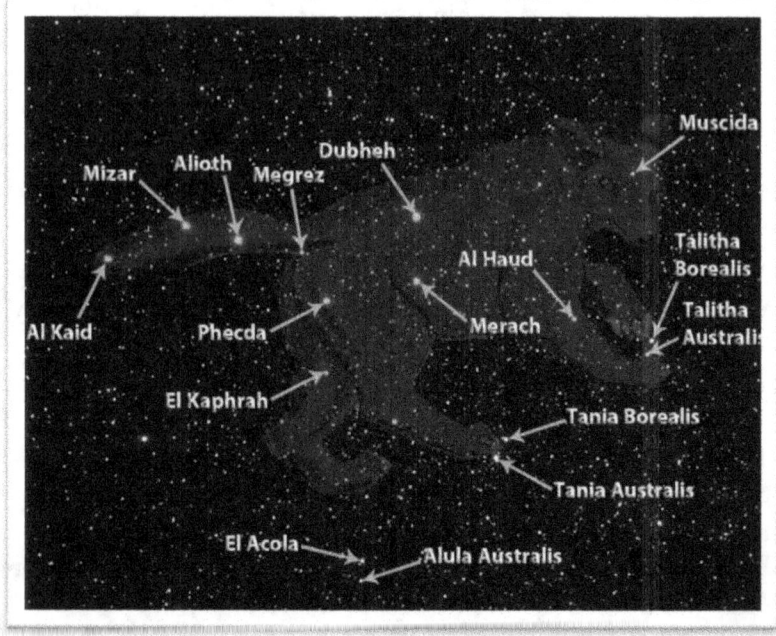

Job 38:32 "Canst thou bring forth Mazzaroth in his season? or canst thou guide Arcturus with his sons?"

Ursa Major is known to most as *the Big Dipper*. It is the Little and Big Dippers that were first known as the sons of Arcturus (Bootes), the Oriental Shepherd. These sons are sheep folds or flocks (Israel and the Church). Although it is represented as a greater bear in Greek and Roman astrology, none of the ancient zodiacs depict such an image. What is depicted in the Little and Great Bears are the sheep folds of God, namely, the *sand* and *star seed* of Israel. After visiting Cancer and seeing the myriads of multitudes of the people of God, here we see yet another image of the same picture. In the Genesis 17:6 blessing and covenant, we see the added statement, "and I will make nations of thee," which is often overlooked. Not only is the Covenant for the sand seed of Abraham prior to Calvary, but also to the star seed after Calvary.

The Star Seed

Romans 4:16 "Therefore it is of faith, that it might be by grace; to the end the promise might be sure to all the seed; not to that only which is of the law, but to that also which is of the faith of Abraham; who is the father of us all." "Now to Abraham and his seed were the promises made. He saith not, And to seeds, as of many; but as of one, And to thy seed, which is Christ" (Gal. 3:16). Paul goes on to explain, "And if ye *be* Christ's, then are ye Abraham's seed, and heirs according to the promise" (Gal. 3:29). Notice below the story behind the stars in this magnificent constellation, of which my personal favorite is *Dubheh Lachar,* meaning "the latter herd, or flock."

THE PRINCIPAL STARS

- *Ash,* (Arabic) meaning "the Assembled."
- *Dubheh,* (Arabic) meaning "herd or flock."
- *El Acola,* (Arabic) meaning "the sheepfold."
- *Cab'd al Asad,* (Arabic) meaning "wealth or multitude."
- *Merach,* (Hebrew) meaning "the flock" (Arab) "the purchased."
- *Mizar,* meaning "separate."
- *El Kaphrah,* meaning "the protected, covered"; (Heb) "ransomed or redeemed."
- *Dubheh Lachar* meaning "**the latter herd, or flock**."
- *Al Kaid,* (Arabic) meaning "the assembled" as in Ursa Minor; (Greek) *Helike* as in Homer's *Iliad* meaning "travelers walking."

CONCLUSION

The stars in the Big Dipper are still speaking to us today. They are telling us of the latter flock, namely the "Church Age." They are telling us that it is not through church membership or good and religious works that we make it into the family of God, but that we must be part of the redeemed of God.[56] They remind us "If ye then be risen with Christ, seek those things which are above, where Christ sitteth on the right hand of God. Set your affection on things above, not on things on the earth. For ye are dead, and your life is hid with Christ in God" (Col. 3:1-3). We are just pilgrims in this life, passing through. This leads us to the third and final decan in Cancer, Argo.

---

[56] Redemption is a doctrine of Salvation that refers to the repurchasing of captured goods; namely, the souls of men from Satan.

Argo "The Pilgrimage"
Chapter 11 Section 4

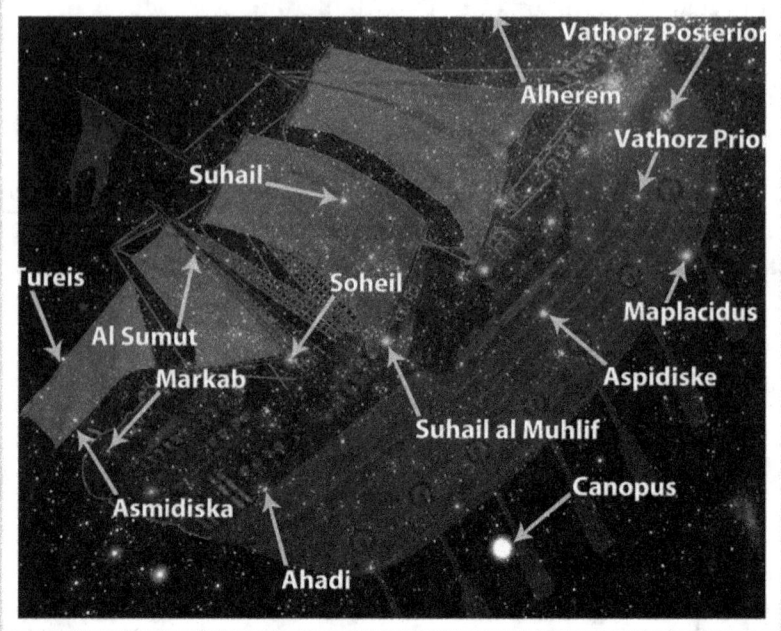

"And the Gentiles shall come to thy light, and kings to the brightness of thy rising. Lift up thine eyes round about, and see: all they gather themselves together, they come to thee: thy sons shall come from far, and thy daughters shall be nursed at *thy* side. Then thou shalt see, and flow together, and thine heart shall fear, and be enlarged; because the abundance of the sea shall be converted unto thee, the forces of the Gentiles shall come unto thee. The multitude of camels shall cover thee, the dromedaries of Midian and Ephah; all they from Sheba shall come: they shall bring gold and incense; and they shall show forth the praises of the LORD. All the flocks of Kedar shall be gathered together unto thee, the rams of Nebaioth shall minister unto thee: they shall come up with acceptance on mine altar, and I will glorify the house of my glory. Who *are* these *that* fly as a cloud, and as the doves to their windows? Surely the isles shall wait for me, and the ships of Tarshish first, to bring thy sons from far, their silver and their gold with them, unto the name of the LORD thy God, and to the Holy One of Israel, because he hath glorified thee." (Isaiah 60:3-9)

In the above Scripture passage, Isaiah predicted the inter-dispensational vision of the Gentiles' engrafting, as he said that they come in on "ships of Tarshish." Likewise, Argo is signifying the people of God in all ages on their spiritual journey to the eternal rest of God.

> By faith Abraham, when he was called to go out into a place which he should after receive for an inheritance, obeyed; and he went out, not knowing whither he went. By faith he sojourned in the land of promise, as *in* a strange country, dwelling in tabernacles with Isaac and Jacob, the heirs with him of the same promise: For he looked for a city which hath foundations, whose builder and maker *is* God. Through faith also Sarah herself received strength to conceive seed, and was delivered of a child when she was past age, because she judged him faithful who had promised. Therefore sprang there even of one, and him as good as dead, *so many* as the stars of the sky in multitude, and as the sand which is by the sea shore innumerable. These all died in faith, not having received the promises, but having seen them afar off, and were persuaded of *them,* and embraced *them,* and confessed that they were strangers and pilgrims on the earth. For they that say such things declare plainly that they seek a country. And truly, if they had been mindful of that *country* from whence they came out, they might have had opportunity to have returned. But now they desire a better *country,* that is, a heavenly: wherefore God is not ashamed to be called their God: for he hath prepared for them a city. (Heb. 11:8-16)

In the passage above, we know from the term used that Abraham dwelt "in tabernacles with Isaac and Jacob, the heirs with him of the same promise," that the signification is that the pilgrimage began with Abraham and Isaac. Jacob continued the pilgrimage into Egypt, which eventually went all the way down into the Church Age. In Argo, we see the wonderful story of the Christian's pilgrimage. In John 17:12-16, Christ taught that we are all pilgrims, not from this world.

"While I was with them in the world, I kept them in thy name: those that thou gavest me I have kept, and none of them is lost, but the son of perdition; that the Scripture might be fulfilled. And now come I to thee; and these things I speak in the world, that they might have my joy fulfilled in themselves. I have given them thy word; and the world hath hated them, because they are not of the world, even as I am not of the world. I pray not that thou shouldest take them out of the world, but that thou shouldest keep them from the evil. They are not of the world, even as I am not of the world." "Neither pray I for these alone, but for them also which shall believe on me through their word." (John 17:12-16, 20)

Christ, speaking of his disciples in these references, included all those who would believe through their word in verse 20, which reaches us in these last days. Let's look at the principal stars in Argo.

THE PRINCIPAL STARS

- *Argo*, (Hebrew) meaning "the company of travelers."
- *Soheil*, (Arabic) meaning "the desired." Indicating the universal church.
- *Asmidiska*, meaning "the released who travel."
- *Canopus*, meaning "the possession."
- *Canobus*, meaning "of him who cometh."
- *Subilon*, (Arabic) meaning "of the branch."

Paul said to the church at Philippi that "our conversation is in heaven; from whence also we look for the Saviour, the Lord Jesus Christ." Conversation in 2016 can mean how we talk to one another, but at the writing of 1611 it meant citizenship. Notice that James Strong said that the underlying Greek meaning of the word is "G4175 πολίτευμα politeuma *pol-it'-yoo-mah* From G4176; a *community*, that is, (abstractly) *citizenship* (figuratively): - conversation."[57] Thus, it is important for Christians to keep their eyes on the "city which hath foundations, whose builder and maker *is* God," not on the world.[58]

---

[57] James Strong, *Strong's Exhaustive Concordance*,

[58] Now therefore ye are no more strangers and foreigners, but fellowcitizens with the saints, and of the household of God; And are built upon the foundation of the apostles and prophets, Jesus Christ himself being the chief corner *stone*; In whom all the building fitly framed together groweth unto an holy temple in the Lord: In whom ye also are builded together for an habitation of God through the Spirit. (Eph. 2:19-22)

Because it is written, Be ye holy; for I am holy. And if ye call on the Father, who without respect of persons judgeth according to every man's work, **pass the time of your sojourning *here* in fear:** Forasmuch as ye know that ye were not redeemed with corruptible things, *as* silver and gold, from your vain conversation *received* by tradition from your fathers; But with the precious blood of Christ, as of a lamb without blemish and without spot: Who verily was foreordained before the foundation of the world, but was manifest in these last times for you. (1st Pet. 1:16-20)

The stars in Argo are still speaking to us today. They are reminding us of the passages of Scripture that admonish us to keep our eyes on the Lord and on our heavenly city. They are admonishing us to not get too attached to the world, but stay focused on our journey and why we are here. John said it well when he said,

Love not the world, neither the things *that are* in the world. If any man love the world, the love of the Father is not in him. For all that *is* in the world, the lust of the flesh, and the lust of the eyes, and the pride of life, is not of the Father, but is of the world. And the world passeth away, and the lust thereof: but he that doeth the will of God abideth forever. (1st Jn. 2:15-17).

THE GOSPEL SHIP

Many moons ago, a ship left the docks of Heaven on a universal journey. That ship has since sailed on many afflicted seas. Its bow is old, its sides are worn, and its sail is patched, but the ship continues to sail eternally. This ship is on a mission with purpose. It can't give up on the voyage, because it has an important rendezvous. Its mission is to pick up all those in the world that will "come unto God" (Heb. 7:25). As it has travelled far and wide, it is making room for "whosoever will" (Rev. 22:17) to come to Christ. It has faced pirates and storms, leviathans and hurricanes, but the ship's Captain has managed to keep everyone safe on the journey.

"For it became him, for whom *are* all things, and by whom *are* all things, in bringing many sons unto glory, to make **the captain** of their salvation perfect through sufferings" (Hebrews 2:10).

"Surely Goodness and Mercy"

A pilgrim was I, and a-wand'ring,
In the cold night of sin I did roam,
When Jesus the kind Shepherd found me,
And now I am on my way home.

Chorus: Surely goodness and mercy shall follow me
All the days, all the days of my life;
Surely goodness and mercy shall follow me
All the days, all the days of my life.
And I shall dwell in the House of the Lord forever,
And I'll feast at the table spread for me;
Surely goodness and mercy shall follow me
All the days, all the days of my life.
All the days, all the days of my life.

He restoreth my soul when I'm weary,
He giveth me strength day by day;
He leads me beside the still waters,
He guards me each step of the way.

(Chorus)

When I walk through the dark, lonesome valley,
My Savior will walk with me there;
And safely His great hand will lead me
To the mansions He's gone to prepare.

(Chorus)

(Lyrics: John W. Peterson and Alfred B. Smith)

LEO "THE LION"
CHAPTER 12 SECTION 1

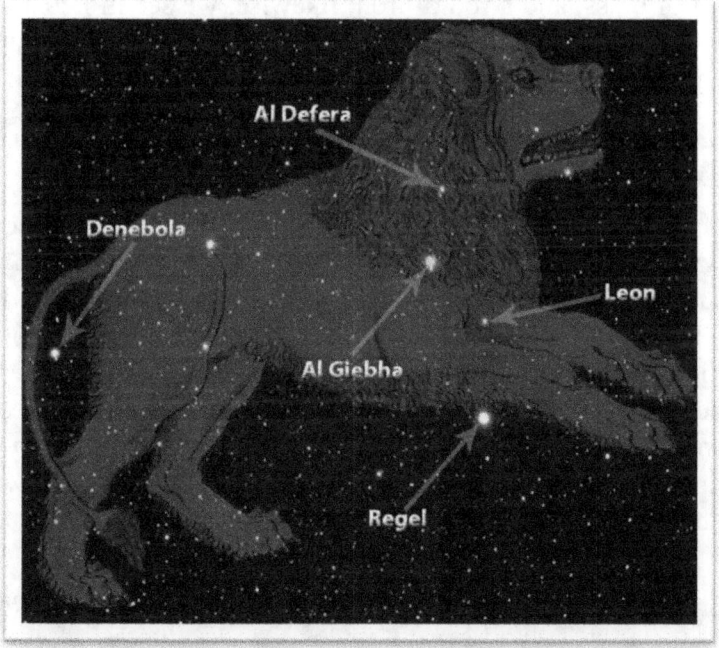

Like the other 11 main constellations, Leo has three decans: *Hydra*, *Crater*, and *Corvus*, all of which present an amazing end to the story of *Biblical Astrology*. In Leo, we come to the last of the main twelve constellations in the Zodiac. Housed in the portico of *the Temple of Esneh* in Egypt there is a great astronomical chart on the ceiling which shows the entire Zodiac with all its constellations. Between the constellations of Leo and Virgo there is an engraving of a Sphinx with a female's head facing Virgo with a Lion's tail pointing to Leo. When this was discovered, it helped astronomers better understand the start and finish of the ancient Zodiac; thus, unlocking one of the greatest mysteries in the Zodiac stories; namely, where does one begin in the cosmic stories? In previous constellations, we have seen many disclosures of the attributes of God and his gospel. In this final *star finale,* we see the Apocalypse in the stars. This new revelation is not as the previous disclosures of a humble and submissive Lamb, but of the powerful and exalted Lion of the tribe of Judah. "And one of the elders saith unto me, Weep not: behold, the Lion of the tribe of Juda, the Root of David, hath prevailed to open the book, and to loose the seven seals thereof" (Revelation 5:5). Like the book of

Revelation, Leo shows the blessed unveiling of God that was not quite revealed in any of the previous constellations. As the Revelation depicts Christ as King of kings, Lord of lords, the preeminent Judge, having eyes of fire, feet of brass, and a rod of iron; likewise, we see in Leo a powerful declaration of the victorious triumph that the Seed of the Woman (Israel) has over the serpent. This leads us to the first principal star in Leo.

PRINCIPAL STARS

Regel is the first principal star in Leo and means "the foot which crushes." This great star is a testament to the greatest and oldest prophecy in the Scriptures found in Genesis 3:15, "And I will put enmity between thee and the woman, and between thy seed and her seed; it shall bruise thy head, and thou shalt bruise his heel." In this prophecy, we see three noteworthy truths. They are: (1) through the seed of the woman would come the One that would bruise the Serpent's head, (2) humanity did not need to be saved from itself, but from its great subjugator and arch enemy of God (Satan), and (3) "the bruising of the heel" would become known to the world as the Sacrifice of Jesus Christ on the cross.

Denebola means "the Judge, the Lord who cometh in haste." "And I saw heaven opened, and behold a white horse; and he that sat upon him was called Faithful and True, and in righteousness he doth judge and make war" (Revelation 19:11). Although we live in a day of much injustice, Denebola enlightens us to the fact that in the last chapter of human history, Christ will reign as the pre-eminent and eternal Judge of righteousness.

Al Giebha means "the exalted." This star is teaching us that one day all the diabolical evils of the world shall come to an end. In order for Christ to be exalted, this must be a prerequisite. Christ today is minimized, belittled, distorted, and mocked by millions around the world. In spite of man's feeble attempts to belittle him, Christ in the end still retains his integrity, virtue, holiness, and power. There is coming a day when the Lamb will be exalted as the Lion. After reading "the Lion of the tribe of Juda," in Revelation 5:5, we see the exalted Lamb in Revelation 5:12-13 "Saying with a loud voice, Worthy is the Lamb that was slain to receive power, and riches, and wisdom, and strength, and honour, and glory, and blessing. And every creature which is in heaven, and on the earth, and under the earth, and such as are in the sea, and all that are in them, heard I saying, Blessing, and honour, and glory, and power, be unto him that sitteth upon the throne, and unto the Lamb for ever and ever." Notice all the attributes that are given to him in the end, which are often minimized in the world, "power, riches, wisdom, strength, honor, glory, and blessing." What a day that will be. Thank God for Al-Giebha's reminder!

> Jesus shall reign where'er the sun,
> Does his successive journeys run;
> His kingdom stretch from shore to shore,
> Till moons shall wax and wane no more.
> (Isaac Watts)

Al Defera means "the putting down of the enemy." This star assures us that Jesus Christ is more powerful than the Devil. Jesus will one day close the last chapter of Satan in two important events. (1) The imprisonment for a thousand years, "And he laid hold on the dragon, that old serpent, which is the Devil, and Satan, and bound him a thousand years" (Revelation 20:2). (2) His eternal damnation to the lake of fire, which is after his final rebellion. "And the devil that deceived them was cast into the lake of fire and brimstone, where the beast and the false prophet are, and shall be tormented day and night forever and ever" (Revelation 20:10).

Leon means "the vehemently coming." This star speaks to us in two ways. First, it is telling us that one day Christ is coming back for his Bride (the church). "Then we which are alive and remain shall be caught up together with them in the clouds, to meet the Lord in the air: and so shall we ever be with the Lord" (1st Thessalonians 4:17). "Caught up" is a Greek compound pronounced in English as *har-pazo* which is worthy of our attention. This signifies that the Bride will be rescued or snatched away from a specific danger. This is assurance that in the time of Satan's free reign with no restraining Holy Spirit (2nd Thessalonians 2:7), the Bride will be out of the picture and will not be permitted to enter the seven year Tribulation mentioned in Daniel and Revelation.[59] At this coming, Christ's feet do not touch the soil of the earth, but instead, the Bride meets Christ in the air. Secondly, Leon is telling us that at the end of the seven year Tribulation period, Christ will defeat Anti-Christ's two hundred

---

[59] Disclaimer: The Holy Spirit is taken away in the sense of God's Kingdom leaving and his purposes with the Church are ended; however, the universal presence of the Holy Spirit are still available during this period.

million man army in the Valley of Megiddo in what scholars call *the Battle of Armageddon* (Revelation 16:16) or *the Second Coming* (Revelation 19)! At this event, Christ's feet touch the earth, splitting the Mount of Olives in half (Zechariah 14:4). In this event, the Bride will return with Christ at the end of the tribulation period, having already gone up with him in the Rapture. "And the armies which were in heaven followed him upon white horses, clothed in fine linen, white and clean" (Revelation 19:14). Other significant stars worthy of observation are:

- *Aryeh*, meaning "he who rends."
- *Al Sad*, meaning "he who tears and lays waste."
- *Pi mentekeon*, meaning "the pourer out of rage."

Conclusion: These stars are still speaking to us today. They are telling us the last chapter in God's gospel. They teach how God finally subdues all the forces of sin and evil. They encourage believers to continue on in dark times by giving us the assurance of the hope that is within us! As we begin to sum up the Apocalypse of the Zodiac, I have one question to ask you dear reader. If Christ were to come back for his Bride today, would you be taken up to meet the Lord in the air? Or would you be left behind to endure the Tribulation Period? Perhaps you don't understand Christ as the Lion because you haven't come to know Christ as the Lamb. The Lamb must precede the Lion. Christ could have called twelve legions of angels to destroy the world and set him free (Matthew 26:53), but it would have ruined his plan to redeem the fallen race. Christ laid down his life as a Lamb to the slaughter (Isaiah 53:7) in order to make the sacrifice that was needed to save the world. That day, Christ who knew no sin, was made sin for us, that we might be made the righteousness of God in him (II Corinthians 5:21). How can we hear such good news and it not tell us how much God really loves us? That's why the Bible says in John 3:16 "For God so loved the world, that he gave his only begotten Son, that whosoever believeth in him should not perish, but have everlasting life."

LEO

Score

Daniel G. McCrillis
Daniel G. McCrillis

HYDRA "THE SERPENT"
CHAPTER 12 SECTION 2

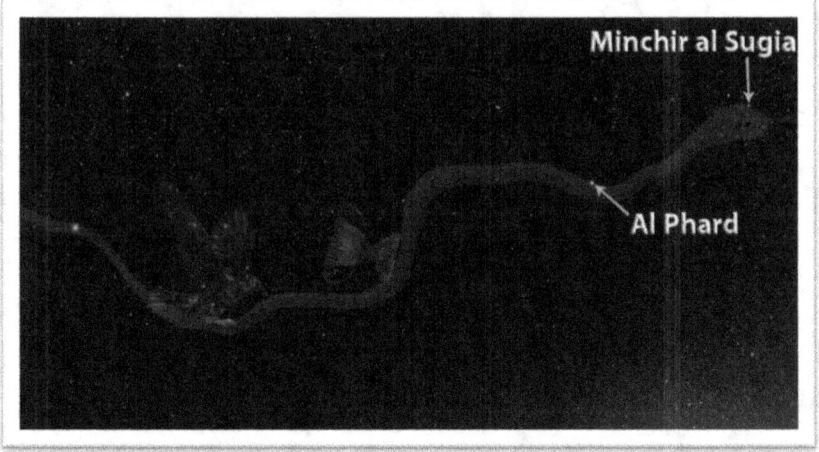

In Leo, we saw the destruction of Satan in the Battle of Armageddon, and in Hydra's stars, we see the destruction of Satan at his *final rebellion,* after the 1,000 year imprisonment! We started off the constellation journey learning that the Promised Seed of the woman would utterly defeat the Serpent. "And I will put enmity between thee and the woman, and between thy seed and her seed; it shall bruise thy head, and thou shalt bruise his heel" (Gen. 3:15). It is here, in the last constellation (Leo) that we see this culmination of the proto-evangelism story. In this image of the arch-enemy of God, we see that he is situated around Leo, Virgo, Cancer and Libra.

THE PRINCIPAL NAME
- *Hydra,* (Heb.) meaning "he is the abhorred."

THE PRINCIPAL STARS
- *Al Phard,* meaning "the separated, excluded, put out of the way."
- *Al Drian,* meaning "the abhorred."
- *Minchir al Sugia,* meaning "the punishing, piercing of the deceiver."

Everywhere in the stars, we have seen the attacks and murderous behavior of the Devil. We see him in Hydra (the serpent), stretching himself around the heavenly equator, as he slithers himself along Libra, Virgo,

Leo, and Cancer. We see him in Libra, because he hates justice. We see him in Virgo, because he was behind "Rachel weeping for her children" (Jer. 31:15; Matt.2:18); for it was Satan who had Herod kill the males, two years old and under. We see him in Leo "the coming one," because he will be there at the Second Coming, in the form of the anti-Christ, having gathered all the nations of the world in the great Battle of Armageddon. Satan is "the lion that walketh about seeking whom he may devour" in the left hand of Orion (Christ). He is Taurus that trophies his congregation of the condemned, while piercing the foot of Auriga (Christ). He is Lepus under Orion's foot. He is Serpens "the serpent" in the hands of Ophiuchus (Christ), searching to get his power and authority. He is Scorpio under the feet of Ophiuchus. He is Cetus the Sea Monster, the great Leviathan of the stars, which attempts to stop the River of Blessing for the believers. He is behind the chaining of Aries and Pisces (the two fishes), because he hates the Jews and the Christian Church. He is the one behind the chains holding Andromeda, the symbol of the Church of God. These chains are his devices (2nd Cor. 2:11) of doubt, confusion, distorted truths, compromises, and sin, with which he has effectively invaded the Church of God. He is Draco, who is after Bootes' (Christ) flocks, namely Israel and the Church (Ursa Major and Ursa Minor). He is the great fighter against all the works of Cygnus, the Northern Cross (the Holy Spirit). Draco is the perversion of godly music (Lyra), the down player of Cepheus the king (Christ), always trying to get people to make Christ take the back seat to him, and he is Centaurus who pierced Lupus "the victim" (picturing Christ) unjustly; signified by the crooked cross.

All over the divine revelation of the stars, we see that there is a Devil attacking and trying to thwart the plan of God. If God was so prevalent to include Satan in these stories, then it is important that we as Christians include him in our teaching and doctrine. We must have a Devil, who had a beginning, who led an angelic rebellion against God, who subjugated the human race, who is presently attacking, who is the originator of all sin, who is constantly working, and is the arch-enemy of the human race and God. His defeat will come one day, but now he is ever destructive and can only be *out-minded* by the Word of God.

Everywhere in the stars, you will see the attacks and murderous behavior of the Devil. You will see him in Hydra (the serpent), which stretches himself around the heavenly equator. He slithers himself along Libra,

Virgo, Leo, and Cancer. You will see him in Libra, because he hates justice. You will see him in Virgo, because he was behind "Rachel weeping for her children." It was he who had Herod kill the males two years old and under. You will see him in Leo "the coming one," because he will be there at the Second Coming in the form of the anti-Christ, having gathered all the nations of the world in the great Battle of Armageddon (Rev. 20:8). Satan is the lion that walketh about in the left hand of Orion (picturing Christ). He is Lepus under Orion's (picturing Christ) foot. He is Serpens "the serpent" in the hands of Ophiuchus (picturing Christ), searching to get his crown back. He's Scorpio under the feet of Ophiuchus (picturing Christ). He is Cetus the sea Monster, the great Leviathan of the stars, which is seen stopping the current of Eridanus (the river of blessing). He is Taurus seen bruising the heel of Auriga, a symbol of Christ and Preserver of the church. He is behind the chaining of Aries and Pisces (the two fish), because he hates the Jews and the Christian Church. He is the one behind the chains holding Andromeda, the symbol of the Church of God. Those chains are the chains of doubt, confusion, distorted truths, compromises, and sin, etc., that he has so effectively invaded the Church of God with. He is Draco, who is after Bootes' (picturing Christ) flocks, namely Israel and the Church (Ursa Major and Ursa Minor). He is the great fighter against all the works of Cygnus the Northern Cross (picturing the Holy Spirit). Draco is the perversion of godly music (Lyra), the down-player of Cephus, the king (picturing Christ), always trying to get people to make Christ take the back seat to him. Lastly, he is Centaurus who pierced Lupus "the victim" (picturing Christ), unjustly, signified by the crooked cross.

CRATER "THE CUP OF WRATH"
CHAPTER 12 SECTION 3

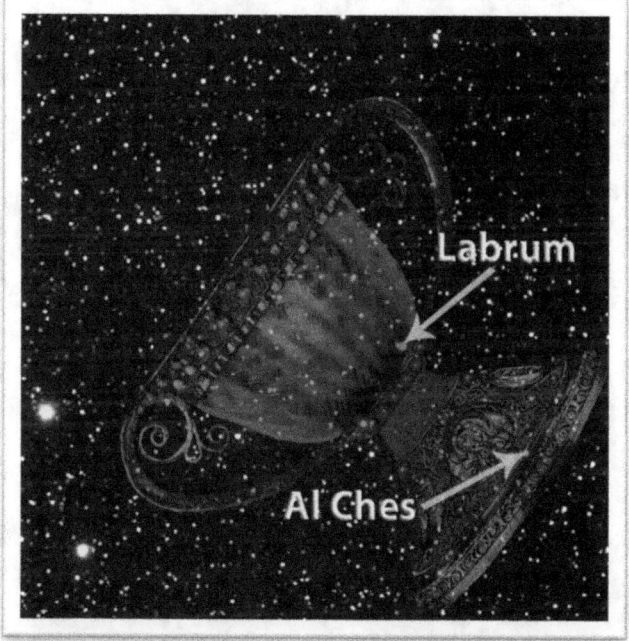

The same shall drink of the wine of the wrath of God, which is poured out without mixture into the cup of his indignation; and he shall be tormented with fire and brimstone in the presence of the holy angels, and in the presence of the Lamb: And the smoke of their torment ascendeth up for ever and ever: and they have no rest day nor night, who worship the beast and his image, and whosoever receiveth the mark of his name. (Rev. 14:10-11)

Crater follows suit with Leo's message, in that it deals with the wrath of God. In this final chapter and second decan of Leo, we see the climax of Satan's future, as God pours out his wrath upon him. "For in the hand of the LORD *there is* a cup, and the wine is red; it is full of mixture; and he poureth out of the same: but the dregs thereof, all the wicked of the earth shall wring *them* out, *and* drink *them*." (Ps. 75:8)

The Principal Stars

- *Al Ches,* meaning "the cup."

157

"And the great city was divided into three parts, and the cities of the nations fell: and great Babylon came in remembrance before God, to give unto her the cup of the wine of the fierceness of his wrath." (Rev. 16:19)

In the gospel of Matthew, Christ prayed to the Father in the Garden of Gethsemane to "let this cup pass from me: nevertheless not as I will, but as thou wilt" (Matt. 26:39), and it is here in the last chapter of the stars that the Father answers his Son's request. It was Isaiah who prophesied all of God's wrath at sin would strike the Messiah. "All we like sheep have gone astray; we have turned everyone to his own way; and the LORD hath laid on him the iniquity of us all" (Isaiah 53:6). At the end of the Astrological story, Satan and sin are finally abolished, there are none others left to redeem; therefore, upon Satan can this cup of fury now be placed. "And the devil that deceived them was cast into the lake of fire and brimstone, where the beast and the false prophet are, and shall be tormented day and night for ever and ever." (Revelation 20:10)

CORVUS "THE BIRD OF PUNISHMENT"
CHAPTER 12 SECTION 4

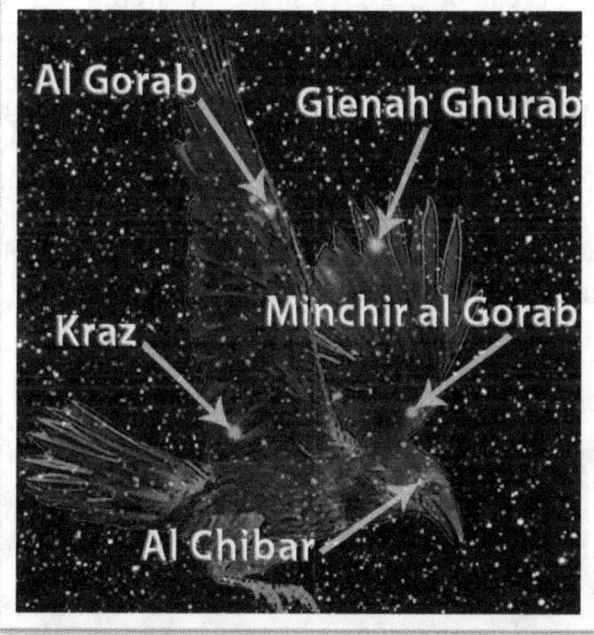

And I saw heaven opened, and behold a white horse; and he that sat upon him *was* called Faithful and True, and in righteousness he doth judge and make war. His eyes *were* as a flame of fire, and on his head *were* many crowns; and he had a name written, that no man knew, but he himself. And he *was* clothed with a vesture dipped in blood: and his name is called The Word of God. And the armies *which were* in heaven followed him upon white horses, clothed in fine linen, white and clean. And out of his mouth goeth a sharp sword, that with it he should smite the nations: and he shall rule them with a rod of iron: and he treadeth the winepress of the fierceness and wrath of Almighty God. And he hath on *his* vesture and on his thigh a name written, KING OF KINGS, AND LORD OF LORDS. And I saw an angel standing in the sun; and he cried with a loud voice, saying to all the fowls that fly in the midst of heaven, Come and gather yourselves together unto the supper of the great God; That ye may eat the flesh of kings, and the flesh of captains, and the flesh of mighty men, and the flesh of horses, and of them that sit on them, and the flesh of all *men, both* free and bond, both small and great. (Rev. 19:11-18)

As we come to the end of Heaven's story, it is worthy to note the parallel between the destruction of Satan and his forces found in Revelation and the star record. In the star record, we see the bird of punishment (Corvus) coming down upon Hydra with his talons and beak; tearing into Hydra's skin. It is at the Lord's Second Coming that we see him coming down from the sky, with wrath, vengeance, fury, and his vesture dipped in blood. Furthermore, in like manner we see *the Final Rebellion* in the Apocalypse, where Satan is loosed out of his 1,000 year confinement for a season. It is here, that we see the destruction of the Serpent.

THE PRINCIPAL STARS

- *Al Gorab*, meaning "the raven."
- *Al Chibar*, meaning "the joining together."
- *Minchir al Gorab*, meaning "the piercing of the raven."
- *Al Chiba*, located in the eye of the raven, means "the curse inflicted."

In the stars of Corvus, we are told of the Raven who destroys the serpent with his beak. In Revelation 19:21 we are told a similar correlation, "And the remnant were slain with the sword of him that sat upon the horse, which *sword* proceeded out of his mouth: and all the fowls were filled with their flesh." Although Satan is not completely destroyed at the Second Coming, we are able to see the destruction of his forces. It will take another thousand years after the Battle of Armageddon before he is finally destroyed.

And I saw an angel come down from heaven, having the key of the bottomless pit and a great chain in his hand. And he laid hold on the dragon, that old serpent, which is the Devil, and Satan, and bound him a thousand years, And cast him into the bottomless pit, and shut him up, and set a seal upon him, that he should deceive the nations no more, till the thousand years should be fulfilled: and after that he must be loosed a little season. And I saw thrones, and they sat upon them, and judgment was given unto them: and *I saw* the souls of them that were beheaded for the witness of Jesus, and for the word of God, and which had not worshiped the beast, neither his image, neither had received *his* mark upon their foreheads, or in their hands; and they lived and reigned with Christ a thousand years. But the rest of the dead lived not again until the thousand years were finished. This *is* the first resurrection. Blessed and holy *is* he that hath part in the first resurrection: on such the second death hath no power, but they shall be priests of God and of Christ, and shall reign with him a thousand years. And when the thousand years are expired, Satan shall be loosed out of his prison, And shall go out to deceive the nations which are in the four quarters of the earth, Gog and Magog, to gather them together to battle: the number of whom *is* as the sand of the sea. And they went up on the breadth of the earth, and compassed the camp of the saints about, and the beloved city: and fire came down from God out of heaven, and devoured them. And the devil that deceived them was cast into the lake of fire and brimstone, where the beast and the false prophet *are,* and shall be tormented day and night forever and ever. (Rev. 20:1-10)

Conclusion to the Constellations

After accepting the challenge to learn and know what the stars teach, we have discovered circumstantial evidence pointing toward the Person and Life of Jesus Christ.

EXTRA CREDITS
THE TABERNACLE ENCAMPMENT

And the LORD spake unto Moses and unto Aaron, saying, Every man of the children of Israel shall pitch by his own standard, with the ensign of their father's house: far off about the tabernacle of the congregation shall they pitch. (Numbers 2:1-2)

The importance and relevance of Numbers chapter two can be seen under the canopy of Biblical Astrology. The Bible informs us that the children of Israel pitched around the Tabernacle and bore an ensign. The word *pitch* is the Hebrew word חנה, pronounced in English as *khaw-naw,* and means to encamp. The word *ensign* is the Hebrew word אות, pronounced as *oth,* and means a flag. The tribes of Israel were to encamp and bare a flag. Next, the Bible informs us that they were to be encamped in a specific order around the Tabernacle.

THE EAST SIDE

And on the east side toward the rising of the sun shall they of the standard of the camp of Judah pitch throughout their armies: and Nahshon the son of Amminadab *shall be* captain of the children of Judah. And his host, and those that were numbered of them, *were* threescore and fourteen thousand and six hundred. And those that do pitch next unto him *shall be* the tribe of Issachar: and Nethaneel the son of Zuar *shall be* captain of the children of Issachar. And his host, and those that were numbered thereof, *were* fifty and four thousand and four hundred. *Then* the tribe of Zebulun: and Eliab the son of Helon *shall be* captain of the children of Zebulun. And his host, and those that were numbered thereof, *were* fifty and seven thousand and four hundred. All that were numbered in the camp of Judah *were* an hundred thousand and fourscore thousand and six thousand and four hundred, throughout their armies. These shall first set forth. (Numbers 2:3-9)

Next unto is represented by the Hebrew word על, pronounced *al,* meaning *over and beyond.* Thus, we have Judah on the south corner of the East Side and Issachar in the north corner of the East Side and Zebulun in-

163

between. You will see this precise pattern throughout the encampment. Furthermore, this is also the exact way that the stars depict it as well.

THE SOUTH SIDE

On the south side *shall be* the standard of the camp of **Reuben** according to their armies: and the captain of the children of Reuben *shall be* Elizur the son of Shedeur. And his host, and those that were numbered thereof, *were* forty and six thousand and five hundred. And those which pitch **by him** *shall be* the tribe of **Simeon**: and the captain of the children of Simeon *shall be* Shelumiel the son of Zurishaddai. And his host, and those that were numbered of them, *were* fifty and nine thousand and three hundred. Then the tribe of **Gad**: and the captain of the sons of Gad *shall be* Eliasaph the son of Reuel. And his host, and those that were numbered of them, *were* forty and five thousand and six hundred and fifty. All that were numbered in the camp of Reuben *were* an hundred thousand and fifty and one thousand and four hundred and fifty, throughout their armies. And they shall set forth in the second rank. (Numbers 2:10-16)

By him is the Hebrew word על, pronounced *al,* meaning *over and beyond.* Thus, we have Reuben on the west corner of the South Side and Simeon on the east corner of the South Side and Gad in-between.

THE MIDST

Then the tabernacle of the congregation shall set forward with the camp of the **Levites** in the midst of the camp: as they encamp, so shall they set forward, every man in his place by their standards. (Numbers 2:17)

The tribe of Levi has no portion, for their portion is the Lord. They pitch but do not bare an ensign. Joseph's son Manasseh takes Joseph's place and Ephraim takes the place of Levi. Thus, there are the two half-tribes Ephraim and Manasseh. One represents Joseph and the other Levi. The idea that Levi is to "set forward" may indicate that they fill gaps, blow trumpets, or possibly both in battle.

On the west side *shall be* the standard of the camp of **Ephraim** according to their armies: and the captain of the sons of Ephraim *shall be* Elishama the son of Ammihud. And his host, and those that were numbered of them, *were* forty thousand and five hundred. And by him *shall be* the tribe of **Manasseh**: and the captain of the children of Manasseh *shall be* Gamaliel the son of Pedahzur. And his host, and those that were numbered of them, *were* thirty and two thousand and two hundred. Then the tribe of **Benjamin**: and the captain of the sons of Benjamin *shall be* Abidan the son of Gideoni. And his host, and those that were numbered of them, *were* thirty and five thousand and four hundred. All that were numbered of the camp of Ephraim *were* an hundred thousand and eight thousand and an hundred, throughout their armies. And they shall go forward in the third rank. (Numbers 2:18-24)

By him is the Hebrew word עַל, pronounced *al,* meaning *over and beyond.* Thus, we have Ephraim on the north corner of the West Side and Manasseh on the south corner of the West Side and Benjamin in-between.

THE NORTH SIDE

The standard of the camp of **Dan** *shall be* on the north side by their armies: and the captain of the children of Dan *shall be* Ahiezer the son of Ammishaddai. And his host, and those that were numbered of them, *were* threescore and two thousand and seven hundred. And those that encamp by him *shall be* the tribe of **Asher**: and the captain of the children of Asher *shall be* Pagiel the son of Ocran. And his host, and those that were numbered of them, *were* forty and one thousand and five hundred. Then the tribe of **Naphtali**: and the captain of the children of Naphtali *shall be* Ahira the son of Enan. And his host, and those that were numbered of them, *were* fifty and three thousand and four hundred. All they that were numbered in the camp of Dan *were* an hundred thousand and fifty and seven thousand and six hundred. They shall go hindmost with their standards. (Numbers 2:25-31)

165

By him is the Hebrew word עַל, pronounced *al,* meaning over and beyond, or opposite corner of the North Side. Thus, we have Dan on the east corner of the North Side and Asher on the west corner of the North Side and Naphtali in-between.

TOTAL AMOUNT

These *are* those which were numbered of the children of Israel by the house of their fathers: all those that were numbered of the camps throughout their hosts *were* <u>six hundred thousand and three thousand and five hundred and fifty</u>. But the Levites were not numbered among the children of Israel; as the LORD commanded Moses. And the children of Israel did according to all that the LORD commanded Moses: so they pitched by their standards, and so they set forward, every one after their families, according to the house of their fathers. (Numbers 2:32-34)

Once this order is seen and equated with the blessings of Jacob in Genesis 49, it is apparent that the positioning of the encampment represented the 12 Constellations of the Heavens and the blessings given by Jacob; in that the ensign (flag) would reflect one the signs of blessing. What are the odds that the encampment would exactly pattern the Zodiac? Once this is done, we can go into the Heavenly Tabernacle mentioned in Psalm 19. The order of the encampment is significant, in that it is patterned after the blessings of Jacob in Genesis 49 and not the order in which Jacob had his children. The blessings of Jacob given to each son represented either one of the twelve constellations or their decans. Thus, in the chart on the following page, we can find: (a) an exact match, (b) relevance to some passages of Scripture which otherwise would have no bearing on anything, and (c) a correlation to several other passages which refer to a pattern of things in heaven (Ex. 25:40; 26:30; 27:8; Num. 8:4; Act. 7:44; and Heb. 8:5). We will further examine these references under the next chapter entitled *the Heavenly Tabernacle.*

In the Old Testament, Christ is introduced by the word ***behold*** four times: (1) "behold the King" (Zech. 9:9); (2) "behold my servant" (Is. 42:1); (3) "behold the man" (Zech. 6:12); and (4) "behold your God" (Is. 40:9). The four gospels depict these very images as well. Matthew depicts Christ as ***King***, Mark as ***Servant***, Luke as ***Man***, and John as ***God***. The early church caught the spirit of the gospels and affixed art work to each of them.

166

Matthew gave Christ the symbol of a *lion*; Mark, the symbol of an *ox*; Luke, the symbol of *man*; and John, the symbol of an *eagle*. These symbols are also used with the cherubim's tetra-morphs in Ezekiel 1:10 and Revelation 4:7.

> As for the likeness of their faces, they four had the face of a man, and the face of a lion, on the right side: and they four had the face of an ox on the left side; they four also had the face of an eagle. (Ezek. 1:10)
> And the first beast *was* like a lion, and the second beast like a calf, and the third beast had a face as a man, and the fourth beast *was* like a flying eagle. (Rev. 4:7)

These images are representative of the four sides of the city of Heaven; each side has three gates made of one pearl. From all sides, one can enter the heavenly tabernacle through the gospel (Pearl of Great Price) of the Lord Jesus Christ. Likewise, Leo represents Christ as the lion of the tribe of Judah and King of Jews (Matthew), Taurus (under his decan Orion) represents Christ as the ox or servant (Mark), Aquarius represents Christ as the son of Man (Luke), and Scorpio, (under the decan of Hercules) represents Christ as the divine Son of God (John). Revelation 21:3 "And I heard a great voice out of heaven saying, Behold, **the tabernacle of God** **_is_ with men**, and he will dwell with them, and they shall be his people, and God himself shall be with them, *and be* their God."

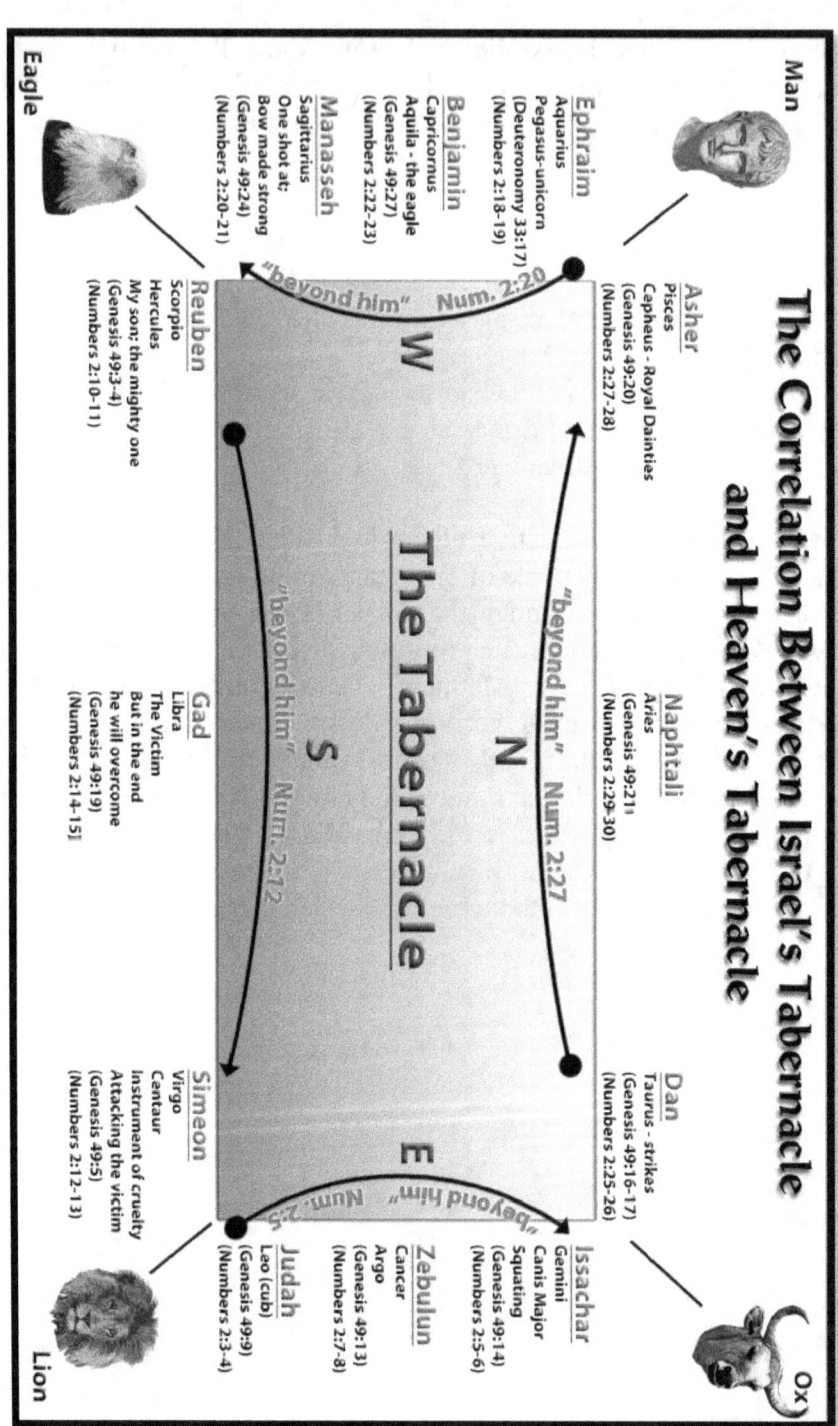

The Correlation Between Israel's Tabernacle and Heaven's Tabernacle

Man

Ephraim
Aquarius
Pegasus-unicorn
(Deuteronomy 33:17)
(Numbers 2:18-19)

Asher
Pisces
Cepheus - Royal Dainties
(Genesis 49:20)
(Numbers 2:27-28)

Naphtali
Aries
(Genesis 49:21)
(Numbers 2:29-30)

Benjamin
Capricornus
Aquila - the eagle
(Genesis 49:27)
(Numbers 2:22-23)

Dan
Taurus - strikes
(Genesis 49:16-17)
(Numbers 2:25-26)

Manasseh
Sagittarius
One shot at:
Bow made strong
(Genesis 49:24)
(Numbers 2:20-21)

Eagle

"beyond him"   Num. 2:20

W

"beyond him"   Num. 2:27

N

The Tabernacle

S

"beyond him"   Num. 2:12

E

Reuben
Scorpio
Hercules
My son; the mighty one
(Genesis 49:3-4)
(Numbers 2:10-11)

Gad
Libra
The Victim
But in the end
he will overcome
(Genesis 49:19)
(Numbers 2:14-15)

Simeon
Virgo
Centaur
Instrument of cruelty
Attacking the victim
(Genesis 49:5)
(Numbers 2:12-13)

Zebulun
Gemini
Canis Major
Squatting
(Genesis 49:14)
(Numbers 2:5-6)

Issachar
Gemini
Canis Major
Squatting
(Genesis 49:14)
(Numbers 2:5-6)

Judah
Leo (cub)
(Genesis 49:9)
(Numbers 2:3-4)

"beyond him"   Num. 2:5

Lion

Ox

THE HEAVENLY TABERNACLE

The 12 sons of Jacob are labeled as stars in Genesis 37:8-10, pictured as the twelve constellations (Genesis 49; Deuteronomy 33) and later their tribes encamped around the earthly tabernacle (Numbers 2:2). According to Hebrews 9:8-12, the earthly tabernacle encampment was a parable of the Heavenly Tabernacle.

> The Holy Ghost this signifying, that the way into the holiest of all was not yet made manifest, while as the first tabernacle was yet standing: Which *was* a figure (parable) for the time then present, in which were offered both gifts and sacrifices, that could not make him that did the service perfect, as pertaining to the conscience; *Which stood* only in meats and drinks, and divers washings, and carnal ordinances, imposed *on them* until the time of reformation. But Christ being come an high priest of good things to come, by a greater and more perfect tabernacle, not made with hands, that is to say, not of this building; Neither by the blood of goats and calves, but by his own blood he entered in once into the holy place, having obtained eternal redemption *for us*. *It was* therefore necessary that the patterns (exhibit for imitation) of things in the heavens should be purified with these; but the heavenly things themselves with better sacrifices than these. (Hebrews 9:8-12, 23)

In Psalm 19:1-4, we see the Heavenly Tabernacle for the sun. This Tabernacle is the Zodiac which God fixed in the heavens.

> The heavens declare the glory of God; and the firmament showeth his handiwork. Day unto day uttereth speech, and night unto night showeth knowledge. *There is* no speech nor language, *where* their voice is not heard. Their line is gone out through all the earth, and their words to the end of the world. In them hath he set a tabernacle for the sun, Which *is* as a bridegroom coming out of his chamber, *and* rejoiceth as a strong man to run a race. His going forth *is* from the end of the heaven, and his circuit unto the ends of it: and there is nothing hid from the heat thereof. (Psalms 19:1-6)

Surprisingly, Philo brought out this teaching in the days of Christ saying,

> When the great High-priest was about to perform the public services enjoined by law, the holy word required that he should in the first place sprinkle himself with water and ashes (Ex. xxix. 4) as a reminder to him of himself... in the next place that he should put on the tunic reaching to the feet, and over it that which he has entitled the embroidered or variegated breastplate (Ex. xxix. 5), a representation and copy of the shining constellations. For there are, as is evident, two temples of God: one of them this universe, in which there is also as High Priest His First-born, the divine Word, and the other the rational soul, whose Priest is the real Man; the outward and visible image of whom is he who offers the prayers and sacrifices handed down from our fathers, to whom it has been committed to wear the aforesaid tunic, which is a copy and replica of the whole heaven, the intention of this being that the universe may join with man in the holy rites and man with the universe.[60]

In Numbers chapter two, we have a linking chapter in the Bible, which links (in type) the earthly tabernacle to that of the heavenly. Notice Numbers 2:2 says, "Every man of the children of Israel shall pitch by his own standard (flag), with the ensign (sign or monument) of their father's house: far off about the tabernacle of the congregation shall they pitch." The ensign of their father's pitch signifies the sign of the patriarchs of Jacob. At the time of Numbers 2 the patriarchs had been dead for quite some time.

The origin of Israel came from God. In the previous chart, there were several reasons why I decided to change the common views of Seiss, Piper, Edersheim and Bullinger and go on my own. It did not make sense to me that God would give us *some* order concerning the patriarchs and the constellations and not *complete* order. If the dots connect, why not connect them? When looking at the constellations and the encampment of the children of Israel, we see that the encampment was divine and in perfect order with the Heavenly Encampment. Let's go through the

---

[60]Philo Vol. V, *On Dreams I*, pg. 411-13, lin. 214-215, Loeb Classical Library, Translated by F.H. Colson and G.H. Whitaker, 1934.

encampment of Israel and look at the blessing given to the twelve patriarchs in Genesis 49 to validate this. Below is a list of the Tabernacle encampment with the affixed constellation of each. What are the odds that the order is the very order of the heavenly constellations?

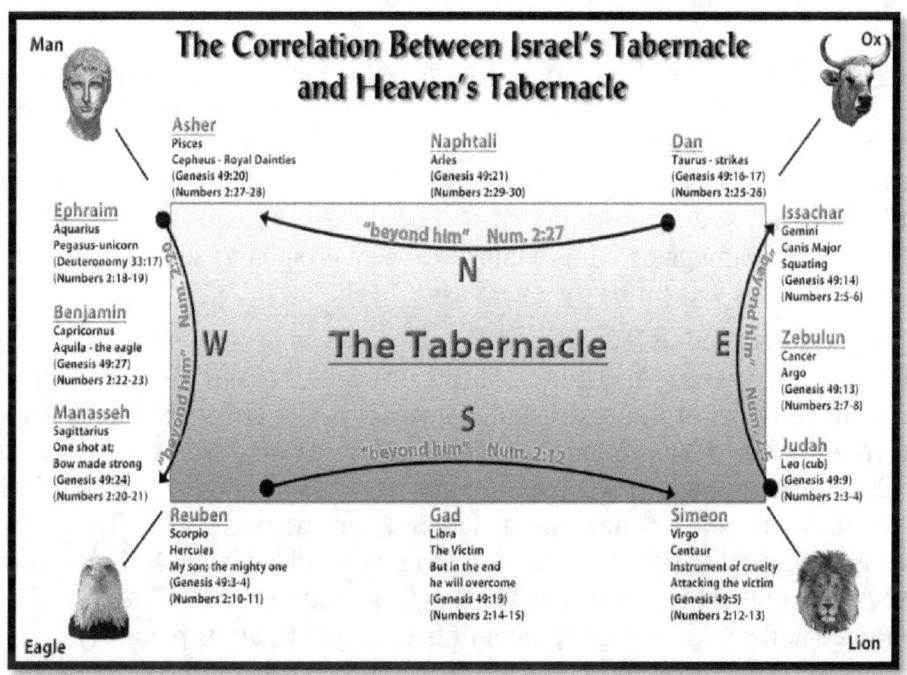

The confusion that arrives by the viewpoint of seeking meaning through the twelve main constellations—without considering their decans— doesn't give us any signification concerning several banners of Israel and their encampment order.

Walking through Jacob's Blessing

When thinking about the order of birth by the Patriarchs of Jacob, we should recognize the following birth order: 1) Reuben Genesis 29:32, 2) Simeon Genesis 29:33, 3) Levi Genesis 29:35, 4) Judah Genesis 29:33, 5) Dan Genesis 30:5-6, 6) Naphtali Genesis 30:7-8, 7) Gad Genesis 30:10-11, 8) Asher Genesis 30:12-13, 9) Issachar Genesis 30:17-18, 10) Zebulun Genesis 30:19-20, 11) Joseph Genesis 30:23-24, and 12) Benjamin Genesis 35:16-18. In Jacob's blessings, there is no specific order of the patriarchs by their birth, but it does agree with Moses' Tabernacle Encampment in Numbers 2, and the 12 Constellations of the Zodiac.

Notice below that the ensign is given through the blessings that Jacob gave to each son. Let's look at these blessings from Genesis 49 and see if we can find precise matches with the constellations.

Judah is represented as Leo and is referenced by "Judah *is* a lion's whelp" (Gen. 49:9).

Zebulun is represented as Argo, a decan of Cancer, and is referenced in "Zebulun shall dwell at the haven of the sea; and he *shall be* for an haven of ships" (Gen. 49:13).

Issachar is represented as Canis Major, a decan of Gemini, and is referenced in "a strong ass couching down between two burdens" (Gen. 49:14). In the image of Canis Major, we see a crouching dog, which in the ancient zodiacs was probably a wild ass.

Dan is represented as Taurus and is referenced in "a serpent by the way, an adder in the path, that biteth the horse heels," (Gen. 49:17) which is seen in the horns of Taurus striking Auriga's heel. The word biteth is the Hebrew word נָשַׁךְ, pronounced in English as *naw-shak'* and means striketh. The thought in the blessing that needs to be conveyed is that Dan was not literally an adder or a serpent, but was as one. These are metaphorical blessings. Furthermore, the word adder is the Hebrew word שְׁפִיפֹן, pronounced in English as *shef-ee-fone'* and means *cerastes* which is a horned adder. All of these points fit the image of Taurus precisely.

Naphtali is represented as Aries and is referenced in "a hind let loose" (Gen. 49:20).

Asher is represented as Cepheus, a decan of Pisces, and is referenced by "he shall yield royal dainties" (Gen. 49:20).

Ephraim is represented as Pegasus, the unicorn, and is referenced by "his horns *are like* the horns of unicorns" (Deut. 33:17). We have to go to Moses account on Ephraim because he was not the firstborn, although he was the preeminent son of Joseph over Manasseh (Gen. 48:20). So, Manasseh is mentioned through the blessing of Joseph and Ephraim is mentioned through Moses.

Benjamin is represented as Aquila, the eagle, a decan of Capricornus, and is referenced in "Benjamin shall ravin *as* a wolf" (Gen. 49:27). Benjamin is not described as a wolf, but his ravening is as a wolf. The word *ravening* is the Hebrew word pronounced in the English as *taw-raf* and means to pluck off or pull. This is done through a bird's beak.

Manasseh is represented as Sagittarius, represented under the blessing of Joseph being the first born (Gen. 48:5-6), "archers have sorely grieved

172

him, and shot *at him*" (Gen. 49:23), "his bow abode in strength, and the arms of his hands were made strong by the hands of the mighty *God* of Jacob" (Gen. 49:24). This is clearly seen in Sagittarius.

Reuben is represented as Hercules, a decan of Scorpio and is referenced by "my might, and the beginning of my strength, the excellency of dignity, and the excellency of power" (Gen. 49:3). This statement conveys the very interpretation of the star names in Hercules.

Gad is represented as the Victim, a decan of Libra and is referenced in "a troop shall overcome him: but he shall overcome at the last" (Gen. 49:19).

Simeon is represented as Centaur, the attacker of the victim and a decan of Virgo, and is referenced by "instruments of cruelty" (Gen. 49:5).

Judah - Leo	Ephraim - Aquarius/Pegasus
Zebulun - Cancer/Argo	Benjamin - Capricornus/Aquila
Issachar - Gemini/Canis Major	Manasseh - Sagittarius
Dan - Taurus	Reuben - Scorpio/Hercules
Naphtali -Aries	Gad - Libra
Asher - Pisces/Cepheus	Simeon - Virgo/Centaur

A FIRST CHRONICLE'S 27 CORRELATION

In addition to the Numbers' chapter 2 encampment and Jacob's blessings in Genesis 49, in 1st Chronicles 27 we have a list of the captains who were over the monthly course of twenty-four thousand men; each captain serving one month in turn (1st Chron. 27:1). The names of the twelve captains and the months in which they served are given in verses 2-15. These men, represented under the twelve tribes, served their king (v. 1) in their given months.

1st Chron. 27:1, Now the children of Israel after their number, *to wit*, the chief fathers and captains of thousands and hundreds, and their officers that **served the king** in any matter of the courses, which came in and went out month by month throughout all the months of the year, of every course *were* twenty and four thousand.

Nīṣān

1st Chron. 27:2-3, "Over the first course for **the first month** *(Niṣañ) was* Jashobeam the son of Zabdiel: and in his course *were* twenty and four thousand. Of the children of Perez *was* the chief of all the captains of the host for **the first month.**" Naphtali

Iyyār

1st Chron. 27:4, "And over the course of **the second month** *(Iyyaî) was* Dodai an Ahohite, and of his course *was* Mikloth also the ruler: in his course likewise *were* twenty and four thousand." Dan

Ṣīwān

1st Chron. 27:5-6, "The third captain of the host for **the third month** *(Siwañ) was* Benaiah the son of Jehoiada, a chief priest: and in his course *were* twenty and four thousand. This *is that* Benaiah, *who was* mighty *among* the thirty, and above the thirty: and in his course *was* Ammizabad his son." Issachar

Tammūz

1st Chron. 27:7, "The fourth *captain* for **the fourth month** *(Tammuî) was* Asahel the brother of Joab, and Zebadiah his son after him: and in his course *were* twenty and four thousand." Zebulun

'Ābh

1st Chron. 27:8, The fifth captain for **the fifth month** *('ABh) was* Shamhuth the Izrahite: and in his course *were* twenty and four thousand. Judah

'Elul

1st Chron. 27:9, "The sixth *captain* for **the sixth month** *('Elul)* was Ira the son of Ikkesh the Tekoite: and in his course *were* twenty and four thousand." Simeon

Tishri?

1st Chron. 27:10, "The seventh *captain* for **the seventh month** *(Tishri?)* was Helez the Pelonite, of the children of Ephraim: and in his course *were* twenty and four thousand." Gad

Marḥeshwaṅ

1st Chron. 27:11, "The eighth *captain* for **the eighth month** *(Marḥeshwaṅ)* was Sibbecai the Hushathite, of the Zarhites: and in his course *were* twenty and four thousand." Reuben

Kiṣlew

1st Chron. 27:12, "The ninth *captain* for **the ninth month** *(Kiṣlew)* was Abiezer the Anetothite, of the Benjamites: and in his course *were* twenty and four thousand." Manasseh

Ṭebheth

1st Chron. 27:13, "The tenth *captain* for **the tenth month** *(Ṭebheth)* was Maharai the Netophathite, of the Zarhites: and in his course *were* twenty and four thousand." Benjamin

Sh^{e}bhaṭ

1st Chron. 27:14, "The eleventh *captain* for **the eleventh month** *(Sh^{e}bhaṭ)* was Benaiah the Pirathonite, of the children of Ephraim: and in his course *were* twenty and four thousand." Ephraim

'Aḍhaṛ

1st Chron. 27:15, "The twelfth *captain* for **the twelfth month** *('Aḍhaṛ)* was Heldai the Netophathite, of Othniel: and in his course *were* twenty and four thousand." Asher

What is significant about the list is that there are twelve captains representing twelve armies, which served their king in their given month. After which, we see the princes of the twelve tribes, in the same chapter. If you take the given facts of the first group and put it in a visual, you have the same pattern and order as **the tabernacle encampment chart** (Numbers 2), the **Abraham Star Seed chart (Genesis 49)**, and **the Zodiac chart**. The Zodiac is the key to unlock the chart, and it is given in Psalm 19:1-6 and Hebrews 9:8-12, 23. This is a fantastic correlation. David would be a type of Christ, the first group of twelve mentioned would represent the star seed of the Abrahamic covenant (the Church), and the second group of twelve would represent the sand covenant (the nation Israel). See chart on following pages. In the law of compound probability, we see a figure of 36 similarities between the monthly, heavenly, and Israel's Tabernacle. The odds of these similarities as they are numerically found are 68,719,476,736 to 1, in favor of a correlation.

STAR SEED CORRELATION

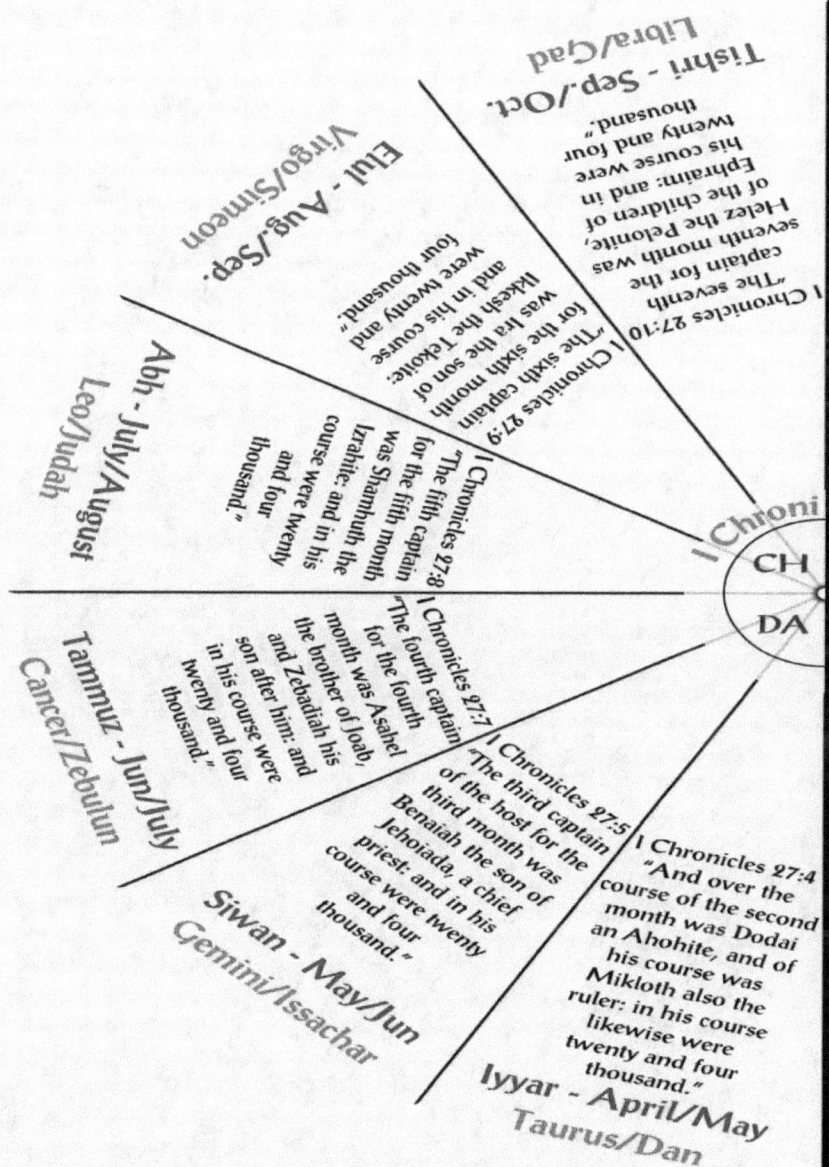

Tishri - Sep./Oct.
Libra/Gad

I Chronicles 27:10 "The seventh captain for the seventh month was Helez the Pelonite, of the children of Ephraim: and in his course were twenty and four thousand."

Elul - Aug./Sep.
Virgo/Simeon

I Chronicles 27:9 "The sixth captain for the sixth month was Ira the son of Ikkesh the Tekoite: and in his course were twenty and four thousand."

Abh - July/August
Leo/Judah

I Chronicles 27:8 "The fifth captain for the fifth month was Shamhuth the Izrahite: and in his course were twenty and four thousand."

I Chronicles 27:7 "The fourth captain for the fourth month was Asahel the brother of Joab, and Zebadiah his son after him: and in his course were twenty and four thousand."

Tammuz - Jun/July
Cancer/Zebulun

I Chronicles 27:5 "The third captain of the host for the third month was Benaiah the son of Jehoiada, a chief priest: and in his course were twenty and four thousand."

I Chronicles 27:4 "And over the course of the second month was Dodai an Ahohite, and of his course was Mikloth also the ruler: in his course likewise were twenty and four thousand."

Siwan - May/Jun
Gemini/Issachar

Iyyar - April/May
Taurus/Dan

I Chroni

CH
DA

I Chronicles 27:1 "Now the children of Israel after of thousands and hundreds, and their officers that came in and went out month by month throughout and four thousand."

WITH I CHRONICLES 27

Marheshwan - Oct./Nov.
Scorpio/Reuben

"The eighth captain for the eighth month was Sibbecai the Hushathite, of the Zarhites: and in his course were twenty and four thousand." I Chronicles 27:11

Kislew - Nov./Dec.
Sagittarius/Manasseh

I Chronicles 27:12 "The ninth captain for the ninth month was Abiezer the Anetothite, of the Benjamites: and in his course were twenty and four thousand."

I Chronicles 27:13 "The tenth captain for the tenth month was Maharai the Netophathite, of the Zarhites: and in his course were twenty and four thousand."

Tebheth - Dec./Jan.
Capricornus/Benjamin

cles 27:1
RIST
SUN
VID

I Chronicles 27:14 "The eleventh captain for the eleventh month was Benaiah the Pirathonite, of the children of Ephraim: and in his course were twenty and four thousand."

Shbhat - Jan./Feb.
Aquarius/Ephraim

I Chronicles 27:15 "The twelfth captain for the twelfth month was Heldai the Netophathite, of Othniel: and in his course were twenty and four thousand."

Adhar - Feb./Mar.
Pisces/Asher

I Chronicles 27:2 "Over the first course for the first month was Jashobeam the son of Zabdiel: and in his course were twenty and four thousand."

Nisan - March/April
Aries/Naphtali

their number, to wit, the chief fathers and captains served the king in any matter of the courses, which all the months of the year, of every course were twenty

178

THE ZODIAC THEORY

In the 1st Chronicles' Correlation, we saw the encampment and a placement of the tribes in their given month, which corresponds precisely to the Zodiac. With that in mind, let's look at Revelation 7:1-8 and gander at a few of the questions it seems to taunt us with. The Zodiac Theory does not claim to be an exact and proper interpretation for all these passages, but a workable model that doesn't dodge most questions in order to answer a few. Probability is what we're after, not possibility or absolute.

> And after these things I saw four angels standing on the four corners of the earth, holding the four winds of the earth, that the wind should not blow on the earth, nor on the sea, nor on any tree. And I saw another angel ascending from the east, having the seal of the living God: and he cried with a loud voice to the four angels, to whom it was given to hurt the earth and the sea, Saying, Hurt not the earth, neither the sea, nor the trees, till we have sealed the servants of our God in their foreheads. And I heard the number of them which were sealed: and there were sealed a hundred and forty and four thousand of all the tribes of the children of Israel. Of the tribe of **Judah** were sealed twelve thousand. Of the tribe of **Reuben** were sealed twelve thousand. Of the tribe of **Gad** were sealed twelve thousand. Of the tribe of **Aser** were sealed twelve thousand. Of the tribe of **Naphtali** were sealed twelve thousand. Of the tribe of **Manasses** were sealed twelve thousand. Of the tribe of **Simeon** were sealed twelve thousand. Of the tribe of **Levi** were sealed twelve thousand. Of the tribe of **Issachar** were sealed twelve thousand. Of the tribe of **Zebulun** were sealed twelve thousand. Of the tribe of **Joseph** were sealed twelve thousand. Of the tribe of **Benjamin** were sealed twelve thousand. (Revelation 7:1-8)

Baptists and Protestants have done a great job in defending the 144,000 position from the Jehovah Witness' view. However, in our estimation, we have not done it justice as far as answering the questions that our stand proposes. Such as: Why exactly 12,000 in number? Will salvation have limitations? How can there be 12,000 out of every tribe? Are the Jews still tallying their genealogies? And why males only? Are women not allowed into God's plan during this time? When considering the four winds," and "the sealing in the foreheads" one might ask many more questions.

Thinking about how there could be twelve thousand souls sealed from each tribe of Israel sometime during the Tribulation Period is a tough one to figure out.

At present, there are no genealogical records being kept for the twelve tribes of Israel. So, how Israel could have 12,000 bloodline males entirely under each tribe is hard, if not impossible, to grasp. Today the twelve tribes are intermingled and inter-raced throughout the Gentile nations; therefore, it is hard to say that there are some Jews, if any, still remaining solely to one particular tribe.

What are the Polls Saying?

How could there be 12,000 from each tribe, when the tribes are sporadic and intermingled? Most people answer this question by saying that "God knows who belongs to what tribe." It can be assumed by this casual approach that one is subconsciously inferring that God will look into the DNA of a particular Jew and see which tribe each individual Jew favors, as to their original tribe, and devote them to that tribe. However, I think the answer to this question is more practical and less mystical than the aforementioned view. Can you imagine if we interpreted the entire Bible with the mindset of "God knows?" So there are four popular views that are held concerning the 144,000. They are:

1) The genealogical records: this view says that there are some forms of genealogy records kept by the Jews that place them to a particular tribe. Having been dispersed for nearly 1,900 years they are intermingled among the world.
2) The Jehovah's Witness view: this view is inter-dispensational, in that it places the bride in the place of tribulation, and links her as the replacement for Israel.
3) The DNA view: this view says that God will place the Jews in the Tribulation into a particular tribe based off of which tribe their DNA leans more in favor of. There are ancestry saliva tests that many are doing now to trace their ancestry, but again we are trying to diagnose tribes within a race.

4) The Last name view: this view says that through their last name, Jews can be placed to a particular tribe.

5) The only way one could understand the end time tribal development being preserved, would be through the Star Seed Covenant of Abraham.

The Zodiac Theory

In the Bible, and with God's elect foreknowledge (the Revelation is prophecy), we see the redemption of the 144,000 servants; in that they were sealed and redeemed (rescued) from the Tribulation. In the Zodiac theory, the sealing took place throughout a year's period of time or one complete zodiac cycle. Throughout a year's period of time, we have four seasons represented by "the four winds" of the earth held by the four angels (7:1).

How could there be twelve thousand sealed from each tribe? The number of people sealed that meet God's criteria in the Tribulation Period during the given zodiacal time slots would accordingly be placed into the spiritual *star seed* of the Abrahamic Covenant and become one of the spiritual tribes of Israel. Therefore, the twelve thousand from each tribe could be more metaphorical or allegorical than literal, thus signifying thousands sealed throughout the given zodiacal months. Since allegoric, symbolic, and metaphoric methods of interpretation are the very language of the Apocalypse, it should at least be considered here, concerning the 144,000 thousand. Furthermore, the

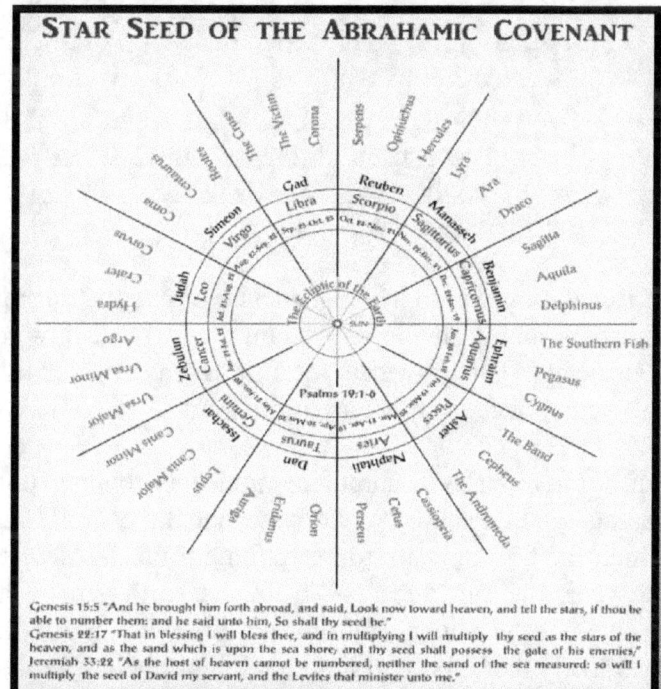

STAR SEED OF THE ABRAHAMIC COVENANT

Genesis 15:5 "And he brought him forth abroad, and said, Look now toward heaven, and tell the stars, if thou be able to number them: and he said unto him, So shall thy seed be."
Genesis 22:17 "That in blessing I will bless thee, and in multiplying I will multiply thy seed as the stars of the heaven, and as the sand which is upon the sea shore; and thy seed shall possess the gate of his enemies."
Jeremiah 33:22 "As the host of heaven cannot be numbered, neither the sand of the sea measured: so will I multiply the seed of David my servant, and the Levites that minister unto me."

181

144,000 are mentioned in chapters 7 and 14. What's significant about these two chapters in the Revelation are that they are *parenthetical chapters* that cannot and should not be placed in a sequence of events.

When one has thoroughly grasped the apocalyptic literature of the Revelation, he can see the probability of the Zodiac Theory. In the following chart we are able to see a visual of the Zodiac Theory, which we have already seen in several places throughout the scriptures.

In the Zodiac Theory, it is my estimation that the tribes, winds and numbers are all allegorical. The allegorical method of interpretation here makes and brings relevance to these obscured passages, which otherwise leave us ever-guessing and wondering. It also magnifies God's sovereignty, in that in God's foreknowledge he has chosen and sealed his elect in the time of anti-Christ. It is also the answer to the martyred saints' request in the fifth seal (see Rev. 6:9-11).

> And when he had opened the fifth seal, I saw under the altar the souls of them that were slain for the word of God, and for the testimony which they held: And they cried with a loud voice, saying, How long, O Lord, holy and true, dost thou not judge and avenge our blood on them that dwell on the earth? And white robes were given unto every one of them; and it was said unto them, that they should rest yet for a little season, until their fellow servants also and their brethren, that should be killed as they *were,* should be fulfilled. (Rev. 6:9-11)

In Revelation we do see a sequence of events, during which the martyred saints (probably of the entire church era) ask how long till God avenges their death. The answer is for a little season (probably the last three and a half years) because there are more martyrs to come.

Furthermore, it is a complete model as to how one can interpret these passages. In Psalm 19, we saw that there is a *Heavenly Tabernacle* in which the Sun appears to be running a race. At certain times throughout the year, the sun appears—from the earth—to be housed in one of the twelve constellations. During the first year of the Tribulation, it is more than likely that God will hold back his angels from hurting the earth or the

182

sea until the elect are sealed from the four corners of the earth. Then, God will begin to pour out his judgments upon the earth. According to this theory, let's say that someone was saved on September 10th. What spiritual tribe would they be grafted into according to our chart? If you said Simeon, you're right. This is a correlative possibility that works. Notice also that the 144,000 enter into **God's Heavenly Temple** in Revelation 7:15! This is unique when thinking of the Zodiac Theory.

> And one of the elders answered, saying unto me, What are these which are arrayed in white robes? and whence came they? And I said unto him, Sir, thou knowest. And he said to me, These are they which came out of great tribulation, and have washed their robes, and made them white in the blood of the Lamb. Therefore are they before the throne of God, and serve him day and night in his temple: and he that sitteth on the throne shall dwell among them. (Rev. 7:13-15)

So, in the star chart on the previous pages, it could be that the 144,000 thousand are sealed over a period of time as the sun travels upon the ecliptic, making the believers during that time a part of the *star seed* of Israel, thus correlating each individual to a specific tribe, and in the thousands. This group will not consist of Jews only, but all those who come under the saving knowledge of Christ. Since "seventy weeks are determined upon thy people and upon thy holy city," (Dan. 9:24) this group is the culmination of both the Old Testament and New Testament saints in the last days. This unique group of believers will picture the millennial saints, where Christ will rule and reign over Jew and Gentile kingdoms.

It is my opinion that they are saved not during the Tribulation Period, but between the Rapture of the Church and the Tribulation.

It is important to know that most commentaries have not embraced the literal interpretation of the 144,000 either. For instance, see what the following commentators had to say about the 144,000. John Gill's Commentary said the following concerning the 144,000:

Rev 7:4 And I heard the number of them which were sealed... And therefore could be sure of the exact number, which did not depend upon his sight, and telling them, in which some mistake might have been made, but he heard the number expressed: *and there were* **sealed an hundred** *and* **forty,** *and* **four thousand**: **which is a square number arising from twelve, the square root of it, being just twelve times twelve thousand; and may denote their being the true and genuine offspring of the twelve apostles of the Lamb, holding their doctrine**, and being built on their foundation; see Revelation 21:14; and these were **of all the tribes of the children of Israel; not that these were all Jews in a literal sense,** for the time of their conversion in great numbers is not yet come. Dr. Goodwin thinks these sealed ones design the believers of the Greek and Armenian churches, and his reasons are not despicable; but this is to limit and restrain them to a particular part of the church of Christ; whereas they take in all the saints within this long tract of time, even all that are the true Israel of God, who are Jews inwardly, **of what nation, kindred, tongue, and people soever**; and is a certain and determinate number for an uncertain and indeterminate one; and only intends a large number of persons known to God and Christ; see the Apocrypha: "Arise up and stand, behold the number of those that be sealed in the feast of the Lord" (2 Esdras 2:38)[61]

See what the People's New Testament said about the 144,000:

Rev 7:4-8 And I heard the number of them which were sealed. The number first named is one hundred and forty-four thousand, twelve thousand from each of the twelve tribes of Israel. **These numbers are not to be taken literally, but only signify that a great number**, not a countless number, but a part of each tribe of Israel, accepted the gospel. Of the tribes Ephraim appears under the name of Joseph, and Dan is entirely omitted, a fact possibly due to the early falling away of Dan into idolatry,

---

[61] *John Gill's Exposition of the Entire Bible*, via, www.e-sword.net.

the number twelve is preserved by counting Levi. For another appearance of the one hundred and forty-four thousand, see Revelation 14:1.[62]

Listen to what Barnes said about the 144,000:

To the other inquiry - whether this refers to those who would be sealed and saved among the Jews or to those in the Christian church - we may answer: (a) **that there are strong reasons for supposing the latter to be the correct opinion.** Long before the time of John all these distinctions of tribe were abolished. The ten tribes had been carried away and scattered in distant lands, never more to be restored; and it cannot be supposed that there was any such literal selection from the twelve tribes as is here spoken of, or any such designation of twelve thousand from each. There was no occasion, either when Jerusalem was destroyed, or at any either time - on which there were such transactions as are here referred to occurring in reference to the children of Israel.

Compare the notes on James 1:1. Accustomed to speak of the people of God as "the twelve tribes of Israel," nothing was more natural than to transfer this language to the church of the Redeemer, and to speak of it in that figurative manner. Accordingly, from the necessity of the case, the language is universally understood to have reference to the Christian church. **Even Professor Stuart, who supposes that the reference is to the siege and destruction of Jerusalem by the Romans, interprets it of the preservation of Christians, and their flight to Pella, beyond Jordan.** Thus interpreted, moreover, it accords with the entire symbolical character of the representation.

(c) The reference to the particular tribes may be a designed allusion to the Christian church as it would be divided into denominations, or known by different names; and the fact that a certain portion would be sealed from every tribe would not be an unfit representation of the fact that a

[62] The People's New Testament, via, www.e-sword.net.

portion of all the various churches or denominations would be sealed and saved. That is, salvation would be confined to no one church or denomination, but among them all there would be found true servants of God. It would be improper to suppose that the division into tribes among the children of Israel was designed to be a type of the sects and denominations in the Christian church, and yet the fact of such a division may not improperly be employed as an illustration of that; for the whole church is made up not of any one denomination alone, but of all who hold the truth combined, as the people of God in ancient times consisted not solely of any one tribe, however large and powerful, but of all combined. Thus understood, the symbol would point to a time when there would be various denominations in the church, and yet with the idea that true friends of God would be found among them all.

(d) Perhaps nothing can be argued from the fact that exactly twelve thousand were selected from each of the tribes. In language so figurative and symbolical as this (Revelation), it could not be maintained that this proves that the santo definite number would be taken from each denomination of Christians. Perhaps all that can be fairly inferred is, that there would be no partiality or preference for one more than another; that there would be no favoritism on account of the tribe or denomination to which anyone belonged; but that the seal would be impressed on all, of any denomination, who had the true spirit of religion. No one would receive the token of the divine favor because he was of the tribe of Judah or Reuben; no one because he belonged to any particular denomination of Christians. Large numbers from every branch of the church would be sealed; none would be sealed because he belonged to one form of external organization rather than to another; none would be excluded because he belonged to any one tribe, if he had the spirit and held the sentiments which made it proper to recognize him as a servant of God. These views seem to me to express the true sense of this passage. **No one can seriously maintain that the writer meant to refer literally to the Jewish people; and if he referred to the**

Christian church, it seems to be to some selection that would be made out of the whole church, in which there would be no favoritism or partiality, and to the fact that, in regard to them, there would be some something which, in the midst of abounding corruption or impending danger, would designate them as the chosen people of God, and would furnish evidence that they would be safe. [63]

Let's see what Vincent Word Studies said:

Not literally, but the number symbolical of fixedness and full completion (12 x 12). The interpretations, as usual, vary greatly, dividing generally into two great classes: one holding that only Jews are meant, the other including the whole number of the elect both Jew and Gentile. Of the former class some regard the sealed as representing Jewish believers chosen out of the literal Israel. Others add to this the idea of these as forming the nucleus of glorified humanity to which the Gentiles are joined. Others again regard them as Jews reserved by God until Antichrist comes, to maintain in the bosom of their nation a true belief in Jehovah and His law, like the seven thousand in the days of Elijah.

The interpretation of the latter class seems entitled to the greater weight. According to the Apocalyptic usage, Jewish terms are "christianized and heightened in their meaning, and the word "Israel" is to be understood of all Christians, the blessed company of all faithful people, the true Israel of God." See Rom. 2:28, Rom. 2:29; Rom. 9:6, Rom. 9:7; Gal. 6:16; Phi. 3:3. The city of God, which includes all believers, is designated by the Jewish name, New Jerusalem. In Rev. 7:3, the sealed are designated generally as *the servants of God*. In chapter 14 the one hundred and forty-four thousand sealed are mentioned after the description of the enemies of Christ, who have reference to the whole Church of Christ; and the mention of the sealed

---

[63] Albert Barnes, *Notes on the Bible Word Studies* via, www.e-sword.net.

is followed by the world-wide harvest and vintage of the earth. The one hundred and forty-four thousand in chapter 14, have the Father's name written in their foreheads; and in Rev. 22:4, *all* the inhabitants of the New Jerusalem are so marked. In Rev. 21:12, the twelve tribes include all believers. The mark of Satan which is in the forehead, is set upon *all* his servants without distinction of race. See Rev. 13:16, Rev. 13:17; Rev. 14:9; Rev. 16:2; Rev. 19:20; Rev. 20:4. The plagues threaten both Jews and Gentiles, as the sealing protects all.[64]

As we can see, the 144,000 have been viewed by several theologians and scholars through the centuries as a cosmopolitan group. As long as we continue to contradict the very method of interpretation intended for the Revelation, we will continue to leave gaping holes in our Bible that leave the student ever learning and never able to come to the knowledge of the truth. Now, I must say, the opposite can be a problem as well. We can read the Bible through the metaphorical eye and catastrophically misinterpret God's Word. Although the first look-through ought to be one of literal interpretation, in several places we must graduate, especially when the Bible gives us permission.

Learning the Song of the Redeemed

And I looked, and, lo, a Lamb stood on the mount Sion, and with him an hundred forty *and* four thousand, having his Father's name written in their foreheads. And I heard a voice from heaven, as the voice of many waters, and as the voice of a great thunder: and I heard the voice of harpers harping with their harps: And they sung as it were a new song before the throne, and before the four beasts, and the elders: and no man could learn that song but the hundred *and* forty *and* four thousand, which were redeemed from the earth. (Revelation 14:1-3)

Something to think about is that there will be a remnant left behind at the Tribulation Period. This is not a remnant of saved, but those who have not reached the age of accountability nor had enough opportunity to receive the gospel of Christ. Our offspring—part of the 144,000—will become

---

[64] Vincent Word Studies via, www.e-sword.net.

"sealed" by God until the day of their redemption. This is signified, in that they will learn a new song.

Redeemed from the earth could possibly indicate that they were sealed sometime after the Rapture, but before the Tribulation. The Christian church and the Jews should be in view here prior to the Tribulation because during the Tribulation, the Jews have little, if no opportunity for a gospel witness during this time. They are deceived during the first half and they are fleeing during the second half. Many have supposed that the two witnesses evangelize the world, but that is hard to find throughout the Revelation. It appears that the ultimate purpose of the two witnesses is to "finish their testimony" (Rev. 11:3-12), not spread the gospel.

The only way one could understand the end time tribal development being preserved, would be through the Star Seed Covenant of Abraham. So the *Tribulation Period* is not a gospel opportunity for the Jew, but a judgment upon Antichrist (False Prophet), the Beast, his Army and the nation of Israel. The *Re-gathering of Israel* will not be fulfilled until the Second Coming and the Millennial Kingdom. In that time it will be a cosmopolitan population of Jew and Gentile, under a hybrid-covenant between Old and New Testament Theology.

It is possible that *the sealing* (7:3-8) of the 144,000 is referring to people left behind that never reached accountability or had a clear presentation of the gospel before the Rapture. These are represented in thousands out of each month of a particular zodiac cycle (7:1). They could also be born again Jews that receive Christ. But the most important thing to notice about this select group is that they join the universal band and brotherhood of Christ for it is in Revelation 7:14, John answered one of the 24 elders saying, "These are they which came out of great tribulation, and have washed their robes, and made them white in the blood of the Lamb." It is significant to note that the final identifying mark of all the heavenly redeemed is seen in Revelation 22:4, "And they shall see his face; and his name *shall be* in their foreheads."

So during the tribulation period there could be thousands of people who come to the knowledge of salvation, that were before the age of accountability or above it but never had a clear presentation of the gospel. They are represented in thousands, saved over a complete zodiacal cycle and are witnesses of the gospel. The Zodiac Theory answers the following questions:

1. Who are the twelve thousand?
2. Why are there twelve thousand exactly?
3. How could there be twelve thousand from each tribe?
4. What did these twelve tribes consist of?
5. What is significant about the four winds?
6. Are the 144,000 saved as we are, or are they delivered?
7. How are the 144,000 saved over time?
8. What does the sealing magnify?
9. How is the sovereignty of God magnified in the Zodiac Theory?
10. What are the most probable methods of interpretation in the Revelation?
11. How can there be a salvation opportunity for the nation of Israel in the Tribulation Period?
12. Is there a freewill involved by those that are sealed?
13. Are the 144,000 saved from sin in the Tribulation or saved from Armageddon?
14. What is significant by 12,000 from each tribe?
15. How are the 144,000 connected to the 5th seal of Revelation chapter 6?

The Zodiac Theory is just a few thoughts of the author. We're not trying to make a declaration of fact, but make propositions to further discover truth. In the end, "let God be true, but every man a liar" (Rom. 3:4).

Future Music Charts

CENTAURUS

Score

Daniel G. McCrillis
Daniel G. McCrillis

Piano

OPHIUCHUS

Score

Daniel G. McCrillis
Daniel G. McCrillis

Piano

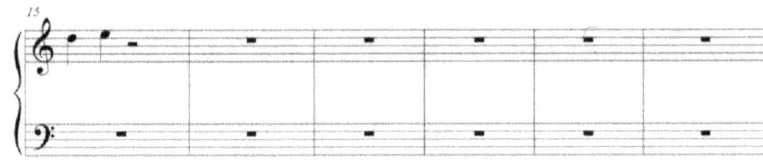

SAGITTARIUS

Score

Daniel G. McCrillis
Daniel G. McCrillis

Piano

LYRA

Score

Daniel G. McCrillis
Daniel G. McCrillis

196

ARA

Daniel G. McCrillis
Daniel G. McCrillis

Score

AQUARIUS

Score

Daniel G. McCrillis
Daniel G. McCrillis

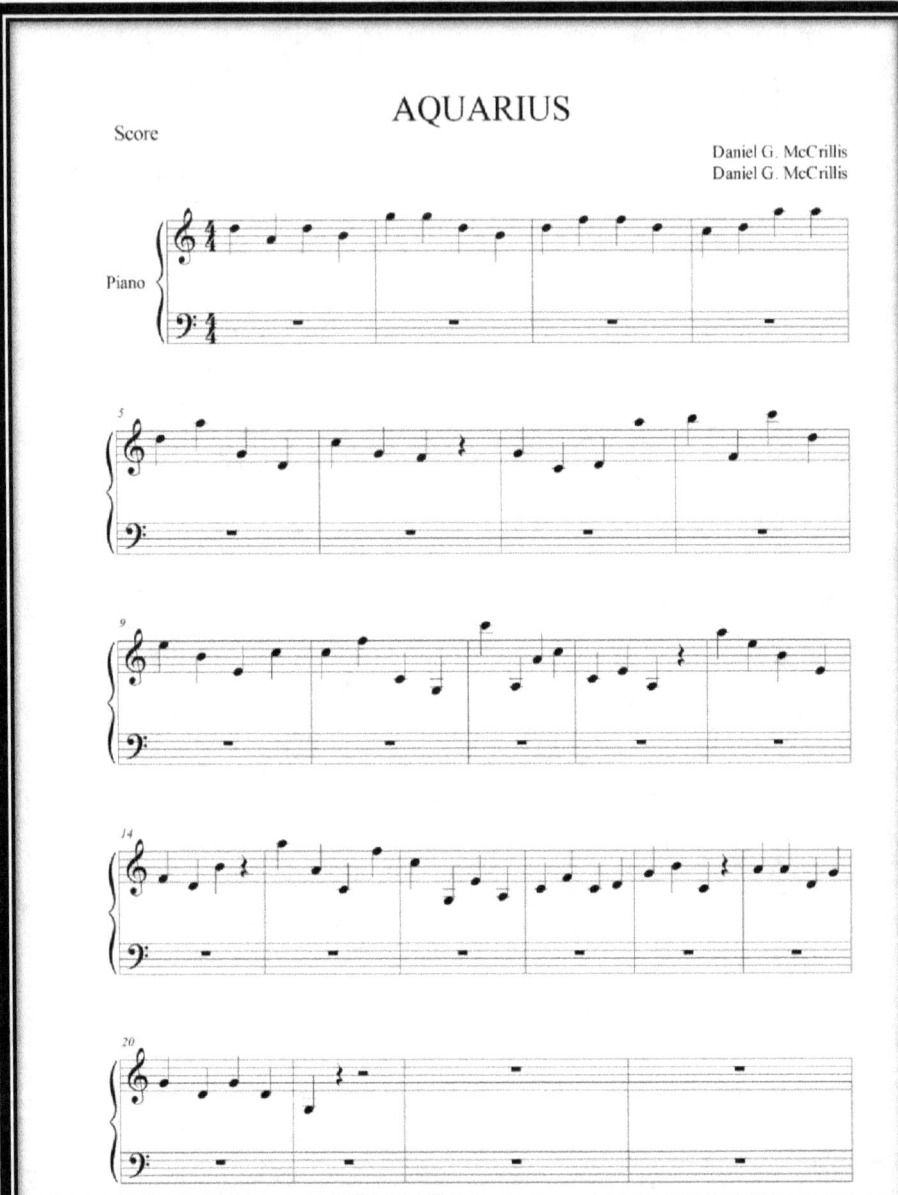

198

PEGASUS

Score

Daniel G. McCrillis
Daniel G. McCrillis

Piano

199

PISCES

Daniel G. McCrillis
Daniel G. McCrillis

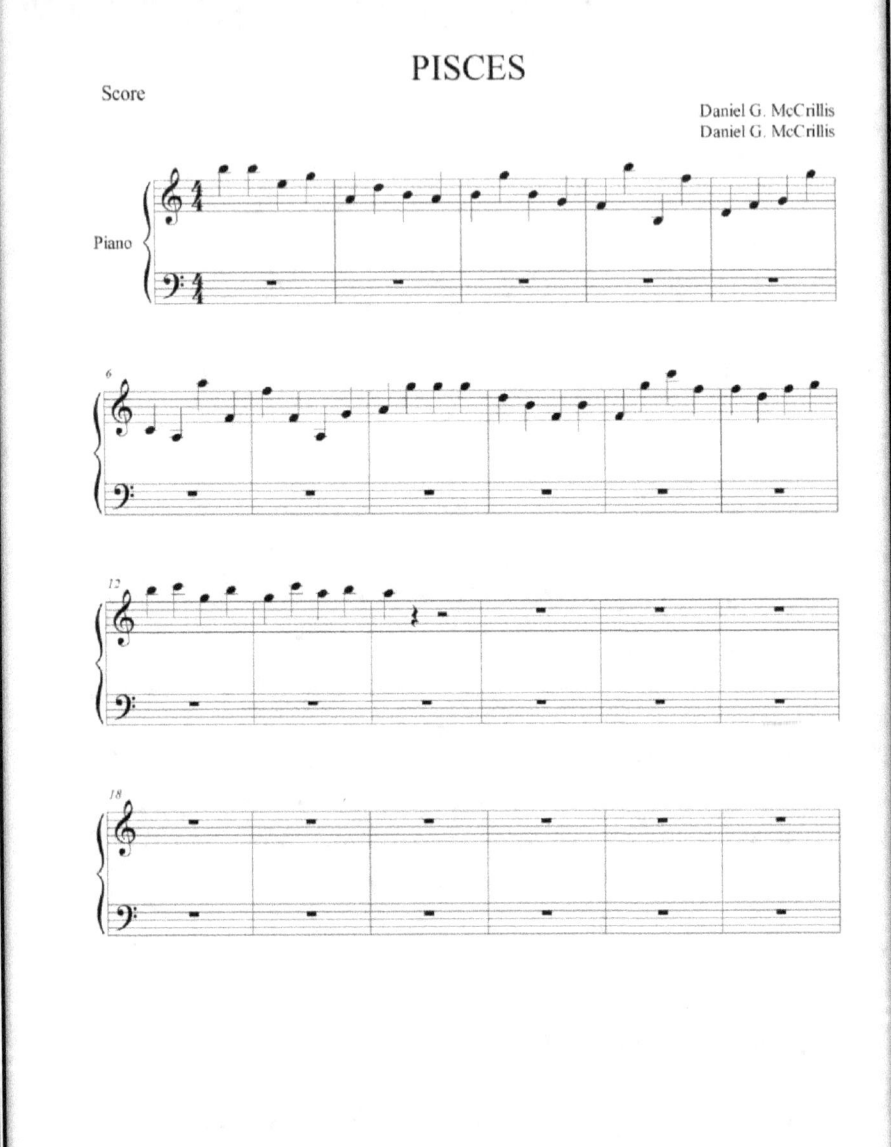

CASSIOPEIA

Daniel G. McCrillis
Daniel G. McCrillis

PERSEUS

Score

Daniel G. McCrillis
Daniel G. McCrillis

TAURUS

Score

Daniel G. McCrillis
Daniel G. McCrillis

Piano

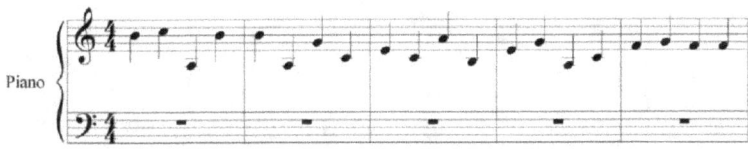

AURIGA

Score

Daniel G. McCrillis
Daniel G. McCrillis

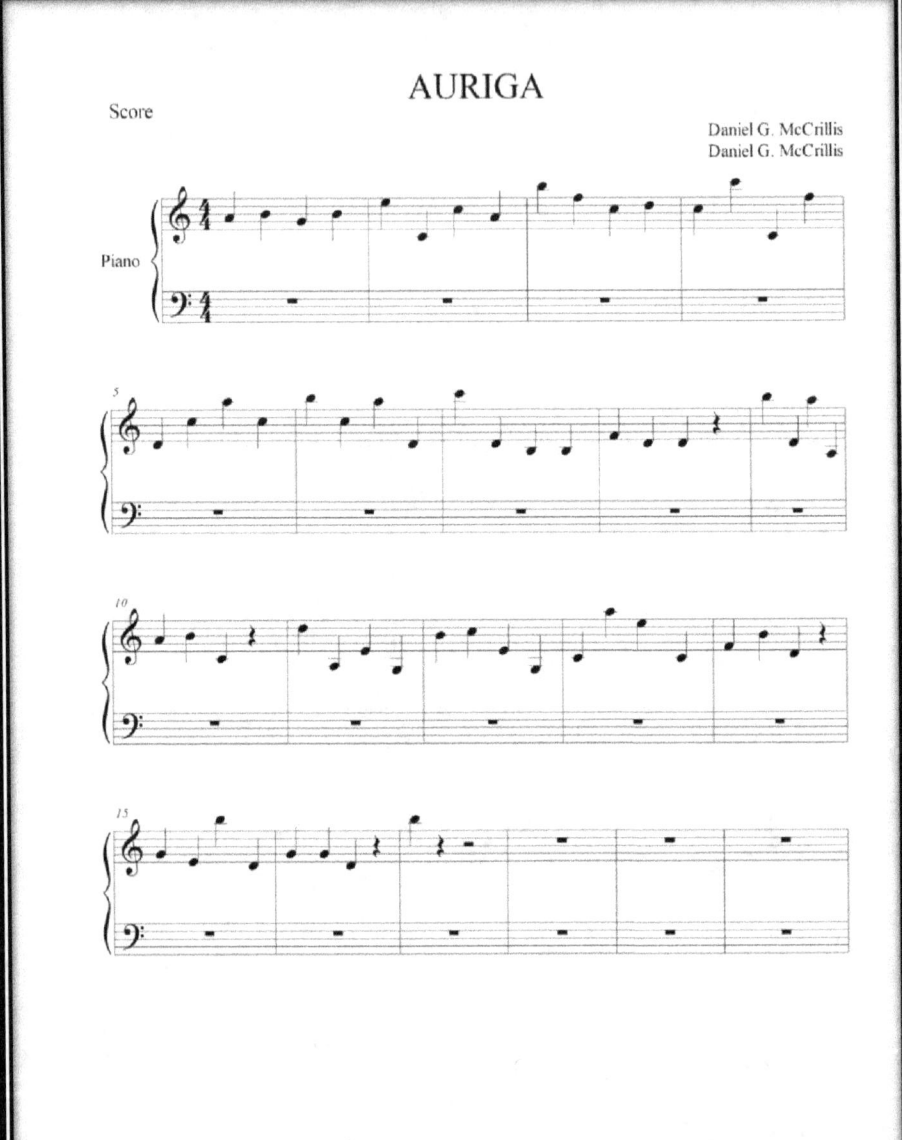

SERPENS CAPUT

Daniel G. McCrillis
Daniel G. McCrillis

Piano

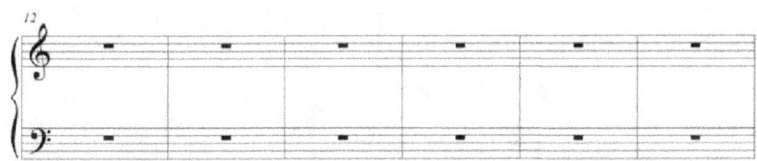

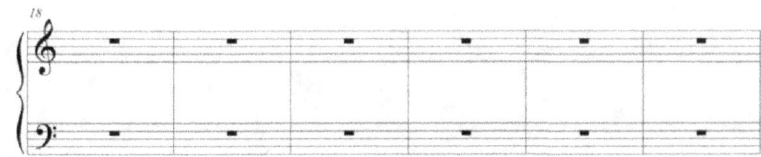

www.ingramcontent.com/pod-product-compliance
Lightning Source LLC
Chambersburg PA
CBHW071256220526
45468CB00001B/156